EYE COLOR, SEX AND RACE

Keys to
Human and Animal Behavior

EYE COLOR, SEX AND RACE

Keys to Human and Animal Behavior

By

Dr. Morgan Worthy

Published By
DROKE HOUSE/HALLUX
116 West Orr Street
Anderson, S. C. 29621

EYE COLOR, SEX AND RACE:
KEYS TO HUMAN AND ANIMAL BEHAVIOR
Copyright ©1974 by Dr. Morgan Worthy

FIRST EDITION

Standard Book Number: 8375-6777-7
Library of Congress Catalog Card Number: 73-86792

Manufactured In The United States of America

Published by
DROKE HOUSE/HALLUX
116 West Orr Street
Anderson, S. C. 29621

Acknowledgements

During the time that I have worked on this project many people have offered me encouragement and to each I am grateful. Three graduate students, Al Markle, Mike Jordan and Richard Price worked very closely with me and each made important contributions to the work.

The manuscript was read and criticized by John McCullers, James Dabbs, Mitchell Morrow, Phillip Wierson and Eileen Sellers. I am indebted to each of them for their interest and helpful comments. I also wish to thank Tina Johns for typing the manuscript.

To my wife and children I express gratitude for their tolerance of my preoccupation with "the book".

Grateful acknowledgement is made to the following for permission to use copyrighted material: Doubleday and Company, Inc., McGraw-Hill Book Company, *Nature,* and W. W. Norton and Company.

201 279

Contents

Part One

Introduction

Chapter 1

Introduction to Eye Color and Behavior

For the last four years I have been obsessed with the fascinating puzzle of what relationship eye color has to behavior. The puzzle has not been completely solved, but many of the pieces are in place and the picture which is emerging is a very interesting one. Far from being meaningless as has been generally assumed, eye color is associated with important behavioral differences.

The Eye Color/Reactivity Relationship

The emerging pattern of associations between eye color and behavior is consistent and may be summarized as follows: *Dark-eyed animals, human and nonhuman, specialize in behaviors that require sensitivity, speed and reactive responses; light-eyed animals, human and nonhuman, specialize in behaviors that require hesitation, inhibition and self-paced responses.*

The specific facts pointing to this conclusion will be discussed and documented in some detail in the following chapters. As an introduction, the details can be briefly summarized by pointing out lines of evidence that converge to present a consistent picture.

(1) *Sports performance.* The studies of eye color done by my students and me were begun more or less accidentally. The research grew out of two casual observations that I made a few years ago. The first was that racial (black/white) differences in sports performance could be organized along a reactive vs. self-paced dimension. Blacks, I observed, tend to perform relatively better in sports that require speed and reactive skills and whites tend to perform relatively better in sports that are not speeded and allow the athlete to perform the task at his own pace. Boxing, for instance, is to a large degree reactive since the boxer must be continually reacting to the moves of his opponent; golf, on the other hand, is mostly self-paced since the golfer can hit the ball when he gets ready. This hypothesis

11

concerning self-paced vs. reactive performance led to quantitative studies of racial differences in various sports.

The second casual observation was that an unusually large number of quarterbacks in professional football were light-eyed. Following that observation, our studies were extended to include not only black/white differences but also differences between dark-eyed whites and light-eyed whites. The findings for racial and eye color comparisons were consistent in that the darker pigmented group in each case tended to perform relatively better on reactive tasks and the lighter pigmented group tended to perform relatively better on self-paced tasks. Wherever significant differences were found they indicated the same pattern. Consequently eye darkness has become the focus of analysis rather than racial or ethnic group differences per se.

Various sports have been systematically studied. In major league baseball the darker-eyed players tend to perform better as hitters than as pitchers; the lighter-eyed players tend to perform better as pitchers than as hitters. Hitting, of course, is primarily a reactive task and pitching is primarily a self-paced task.

Basketball is largely a reactive sport and dark-eyed players tend to do very well in this sport. However, light-eyed players tend to excel at shooting free throws which is the one aspect of the game that is completely self-paced.

Football is a sport in which there are very interesting racial and eye color differences. In professional football the greater the proportion of blacks playing a particular position (e.g., defensive line) the darker, on the average, the eyes of the whites who play that position. The position with the greatest percentage of blacks is defensive back; white defensive backs as a group are darker-eyed than white players in any other position. The position with the lowest percentage of blacks is quarterback; white quarterbacks are lighter-eyed, on the average, than white players in any other position. Of the various football positions, defensive back is probably most dependent on speed and reactive ability and quarterback is probably least dependent on those skills.

There are other bits of evidence from sports such as bowling, golf, and sprinting to suggest that dark-eyed athletes tend to perform relatively better in sports that require speed and reaction and light-eyed athletes tend to perform relatively better in sports that allow self-paced or nonreactive performance.

(2) *Automatic response.* Dark-eyed people appear, in general, to be more physiologically and behaviorally responsive to various physical

12

stimuli than light-eyed people. Various studies of ethnic differences in pain reaction have compared whites with blacks, whites with Orientals, Irish with Italian, etc. In these studies it has been repeatedly found that dark-eyed ethnic groups exhibit stronger pain reactions than do light-eyed ethnic groups. In addition to the study of ethnic groups, a study of Caucasian patients done in Australia at a dental clinic found that the darker the eyes of the patient the greater the pain reactions exhibited during dental drilling.

Other evidence, also, indicates that dark-eyed people are more physiologically sensitive or reactive than are light-eyed people. Their lower thresholds of physiological excitation is consistent with skill in reactive tasks that require sensitivity and responsiveness to the physical environment.

(3) *Perceptual responses.* Eye color appears to have an effect on how one sees the world. For certain visual functions, such as visual acuity in bright light, dark eyes are superior to light eyes. Animals with light eyes are bothered by strong sunlight, but they may have a visual advantage under other conditions.

Another important aspect of visual perception is the perceptual competition between response to color cues versus response to form cues. Since color is usually the dominant or first noticed cue, psychologists have traditionally considered the tendency to respond primarily to color as indicative of an uninhibited or emotionally reactive response. Since attending to form often requires inhibition of an immediate response to color, the tendency to respond primarily to form has been considered indicative of behavioral inhibition or emotional nonreactivity. Researchers have found that people with dark eyes respond relatively more to color and people with light eyes respond relatively more to form. These findings provide strong support for the generality of the eye color/reactivity relationship in areas other than sports or physical performance.

Not only do light-eyed and dark-eyed animals differ in the response to color as such, but they also differ in perception as a function of the particular color involved. Dark-eyed animals—both human and nonhuman—are particularly responsive to colors at the long wave (red and yellow) end of the color spectrum. Short wave light is partially blocked by heavy pigmentation, and, perhaps for that reason, animals with dark eyes are less responsive to colors at the blue end of the color spectrum than are light eyed animals. Among humans, it has been reported that Northern Europeans prefer blues and grays rather than the reds and yellows preferred by Southern Europeans. This difference has been traditionally attribu-

13

ted to cultural influences, but it may, in fact, be genetic in origin.

Perceptual tendencies tend to some degree to influence the patterns of cognitive strengths and weaknesses among humans. Dark-eyed groups tend to do better on cognitive tasks that require perceptual speed and accuracy than on cognitive tasks that involve spatial relations. Light-eyed groups tend, on the other hand, to do relatively better on the spatial tasks. It is important to note that what is referred to here is differences in patterns of cognitive abilities rather than differences in overall cognitive ability. Studies of light-eyed Caucasians and dark-eyed Caucasians (done in the 1920's) found that the two groups did not differ in overall I.Q.

(4) *Human social responses.* This line of evidence is more sketchy and tangential than are the others, but again what is known is consistent with the overall picture. In any human environment, other people comprise a strong stimulus to which we may be more or less responsive. Just as we can think of color as usually a strong and dominant cue in the perceptual field, we can also think of other people as a strong or dominant cue in the environmental field. Groups and individuals differ in the degree to which their behavior reflects reaction to these personal or social cues.

Northern Europeans are typically thought of by other people as cold or distant; privacy is maintained and valued. Southern Europeans and other dark-eyed groups are more responsive to other people; less social distance is maintained and there is more physical contact. These differences in social responses are attributed to differences in cultural learning. However, it is reasonable to assume that cultural practices evolved consistent with the genetic predispositions of the population in question. Making that assumption, it appears that dark-eyed human groups are more predisposed to social responding than are light-eyed human groups.

This conclusion is bolstered by some scattered within-culture research findings based on studies of light-eyed and dark-eyed Americans. It has been demonstrated that dark-eyed males are more responsive to, or influenced by, the opinions of another person than are light-eyed males. There is, also, a possibility that dark-eyed men are particularly effective in leadership positions that require personal or socio-emotional leadership ability.

As noted earlier the findings based on social responses are more exploratory and less conclusive than the other lines of evidence, but, to the degree that there is information, it is consistent with the overall picture.

(5) *Nonhuman animal behavior.* Studies of nonhuman animal

behavior provide the most convincing line of evidence to support the idea that the eye color/reactivity relationship is a general phenomenon in nature and is genetic in origin. We can see from animal studies that the reactive-nonreactive dimension involves important differences in natural adaptation to varying environmental demands.

Many species of birds eat insect or animal life; to the degree that they do so, they are hunters and must use some type of hunting tactic to capture the prey. One tactic, which is reactive in nature, is to immediately pursue the prey as soon as it is sighted. Many species of birds, almost all dark-eyed, use this tactic to catch insects "on the wing". Light-eyed birds that eat insects are more likely to hunt on the ground or in other situations in which speed and reactive skills are less important.

A similar difference is found among birds of prey that hunt other birds for food. Falcons, mostly dark-eyed, are very speedy and catch birds on the wing by means of direct attack. Many light-eyed species of hawks also feed on birds, but they, in contrast, are slower and depend on various methods of surprise and ambush. The methods used by light-eyed hawks involve delay and inhibition of immediate response.

Likewise, birds and mammals that stalk their prey are primarily light-eyed. In this regard, cats are the prime example, but there are other light-eyed examples, such as wolves and herons. The tactics differ in their specifics from one example to the next, but what is common to all is the use of indirect approach or delay of attack until the advantage is with the predator. During the period of delay the predator may actively stalk the prey or the predator may simply remain still and wait for the prey to come closer. In all cases, the success of the tactic depends on the inhibition of direct and immediate attack.

Just as there are eye color differences associated with different hunting tactics employed by predators there are also eye color differences associated with different escape tactics employed by the prey. Among birds, most species escape from predators by means of reactive flight; dark-eyed birds specialize in that method of escape. Some species of birds escape by means of "freezing" or some method of concealment; light-eyed species are more likely to engage in these nonreactive escape tactics than are dark-eyed species.

The examples that have been given involve differences between species, but differences of this nature can also be found within a species. Domesticated dogs comprise one species, but there are large differences between breeds (races) in behavior. Most breeds are dark-eyed but among the breeds of hunting dogs selectively bred for ability to freeze in response to game (i.e., pointers and setters), most breeds are light-eyed.

Laboratory studies of rodents have also found a tendency for light-eyed strains to have a predisposition for freezing behavior. This difference, coupled with the tendency of dark-eyed strains to be behaviorally reactive, helps to explain why the light-eyed strains are superior on some types of learning tasks and the dark-eyed strains are superior on other types of learning tasks.

There are many gaps in our knowledge of eye color and reactive behavior, but when one considers the number and variety of converging facts supporting the same conclusion, it can not be easily discounted or ignored. Even so, the conclusion drawn may seem bizarre to some, since eye color appears to be so far removed from behavior. Actually, in the context of what is known in the fields of physiology and genetics, the connection between eye color and behavior is not at all bizarre. That is not to say that the physiological and genetic links between eye color and behavior have been positively demonstrated, but only that such links are plausible in the context of present knowledge.

As everyone is aware, the eye has important visual functions, but it also has important nonvisual functions. In addition to the nerve tracts that go from the eye to the visual cortex, there are also smaller tracts that go to other centers in the brain—especially the pineal gland and the hypothalamus. These centers are responsive to light which is mediated by the eye as well as to light which directly penetrates the skull. The response of these organs to light is wave length dependent, which is an important consideration since pigmentation in the eye, as elsewhere, blocks more short wave light than long wave light.

The pineal and the hypothalamus serve as regulators of the various endocrine glands. These glands are involved in many functions, including bodily activation or excitation. There is evidence that long wave light facilitates activation and short wave light inhibits activation. There is a chain of effects, then, which potentially connects eye pigmentation to those physiological functions that mediate differences in behavioral reactivity.

Human eye color depends primarily on the amount of brown pigment, melanin, present in the various layers of the iris. If the granules of melanin are very small there is an optical effect which makes the eye look blue. With progressively larger granules of melanin the eye appears progressively darker: gray, green, hazel,

16

brown, black. Other colors such as yellow or red may be observed in animal eyes. With humans or animals, irises that are plain brown or black may, for convenience, be categorized as dark and all other colors categorized as light.

The amount of pigmentation found in the eye is genetically determined and depends on several different genes (i.e., is polygenic). One of the genes that helps determine eye pigmentation is sex-linked. Without getting into the specifics of what this means genetically, it is noteworthy that females have slightly darker eyes than do males. It is also an interesting fact that on many measures females are more reactive than are males.

Implications For An Evolutionary Psychology

The eye color/reactivity analysis presented here attempts to integrate numerous and varied facts into a meaningful theoretical generalization. It relies not on the certainty of any one fact, but on the convergence of many facts. In this regard and in general orientation the analysis is an attempt to utilize the Darwinian model.

The Darwinian theory revolutionized biology, but it has not had a comparable influence on modern psychology. Although attention is paid to the behavioral genetics of animal behavior, there is often resistance to genetic interpretations of human behavior. Psychological theories often tend to treat human individuals and groups as interchangeable parts varying only to the degree that cultural learning differs.

This overemphasis on learning has been so dominant a force, that psychology has not fully benefitted from the Darwinian break-through. This has happened despite the fact that Darwin himself considered behavioral characteristics no less a product of evolutionary forces than physical characteristics. Two emphases of the Darwinian theory have been particularly ignored by psychologists. The first is the similarity between humans and other species; the second is the great variability within each species. Those two points are readily accepted when applied to morphological analyses, but often emotionally resisted when applied to behavioral analyses.

There is some evidence, though, that behavioral scientists are becoming more aware of biological considerations. Several popular books in recent years have attempted to place human behavior in an evolutionary framework. More and more behavioral scientists are being forced to share the view of the noted anthropologist, Edward T. Hall, stated in the preface to his 1966 book, *The Hidden*

Dimension:

As an anthropologist I have become accustomed to going back to the beginning and searching out the biological substructures from which a given aspect of human behavior springs. This approach underscores the fact that man is first, last, and always, like other members of the animal kingdom, a prisoner of his biological organism. The gulf that separates him from the rest of the animal kingdom is not nearly so great as most people think. The more we learn about animals and the intricate adaptation mechanisms evolution has produced, the more relevant these studies become to the solution of some of the more baffling human problems (p. ix-x).

This need for an approach based on sound biological principles has also been pointed out by Nicholas Thompson of Clark University. Writing in a recent issue of the *American Psychologist,* he emphasized the value of a broad, interdisciplinary, behavioral science, which he referred to as "psychobiology".

Psychobiology (or sociobiology) is the study of human behavior from a biological perspective. A biological perspective sees not only the machinery of the individual organism but also the forces acting upon whole populations of organisms which results in their adaptation and evolution. Psychobiology is not yet itself an organized discipline, but it is an area of concern to which many disciplines have made contributions. Anthropology has contributed information about past evolution and present variation in primate and human morphology and behavior; psychology has produced a body of evidence on the development of behavior and the physiological control of behavior in a limited number of standard laboratory subjects; and zoology has developed the theories of behavior evolution and a large body of facts concerning the social systems of animals (Thompson, 1972, p. 581).

I am hopeful that this book will play some part in the continued emergence of evolutionary psychology or psychobiology. The problems of mankind demand that we seek to understand behavior; most of our survival problems are basically problems of behavior.

Social Implications

Perhaps some would question whether studies of genetic differ-

18

ences in behavior among humans should be done because of possible misuse of the information. My faith in the matter is that increased knowledge always leads ultimately to more good than harm.

Of the three graduate students who have worked closely with me on this research project, one has brown eyes, one has blue eyes and one has hazel eyes. As a sequel to this account, I have suggested that they should each write monographs entitled respectively: "The natural superiority of dark eyes", "The natural superiority of light eyes" and "In all things moderation or the best of both worlds". The suggestion was made in jest, but it is probably true that a case could be made for each of the three positions by selective emphasis. In fact, dark eyes are not superior to light eyes in any absolute sense nor is the converse of that true. Neither can reactivity nor nonreactivity be considered as superior or more desirable in any absolute sense. There is always the contrast of what is an advantage in one situation being a disadvantage in another situation.

There is, of course, a tendency for different human groups and individuals to place heaviest evaluative weight on their own areas of special skill. In moderation that may have a positive effect on self esteem, but we should keep in mind the often arbitrary nature of such value judgments. If we are to truly appreciate and respect talents different from our own, we must recognize the degree to which we are enriched by individual and group diversity. In our evaluations of what is good or important, we must resist the temptation to load the evaluative dice with those skills on which we excel. By taking a broad view of human abilities, we easily and honestly reject the claims of generalized superiority or generalized inferiority.

Many behavioral scientists are reluctant to seriously and objectively consider human genetic differences because of the fear that such findings will undermine feelings of brotherhood and equal worth. Such fears may be justified to some degree, but they reflect the outmoded belief that respect can follow only from perceived similarity. That assumption has, on the whole, been rejected in regard to individual and group differences other than those that involve genetic differences in behavior. In religious matters and other cultural areas of concern, we no longer expect or demand similarity, but have developed instead a philosophy of enlightened pluralism. This philosophy leads us to accept and respect people whose beliefs and life styles are different from our own. If, in the future, we are to increase feelings of human brotherhood we must expand the pluralistic philosophy to include genetic, as well as cultural, differences in behavior.

Organization Of The Book

The book is organized into four parts; this introductory chapter comprises Part One.

Part Two, consisting of Chapters 2 through 6, deals with eye color as it relates to speed and reaction in gross motor behavior.

Part Three, consisting of Chapters 7 through 10, deals with the eye and its relationship to physiological functioning.

Part Four, consisting of Chapters 11 through 13, discusses the idea of relationship of eye color to a variety of variables such as social behavior, physical size and various physical disorders. An interesting parallel between eye color differences in behavior and sex differences in behavior is also discussed.

For ease of reading, bibliographical references and statistical details are placed in parentheses so that they can be easily skipped by the non-technical reader.

The book, as a whole, summarizes the results of my efforts to understand the puzzle of eye color and behavior. Because it is to some degree a personal account of research, I have tried to maintain some of the chronological order. Hopefully, this approach will allow the reader to sense some of the challenge and excitement that goes with trying to fit together the pieces of a scientific puzzle.

References

Hall, E. T. *The hidden dimension.* Garden City, New York: Doubleday and Company, 1966.

Thompson, N. S. Psychobiology as a form of general education. *American Psychologist,* 1972, *27,* 580-582.

Part Two

Eye Color and Behavioral Reactivity

Chapter 2

Race, Eye Color and Self-Paced
vs. Reactive Sports Performance

Several years ago a series of studies were begun that initially had nothing to do with eye color as such. They were concerned with racial differences in sports participation and performance. Based on casual observation I had noticed that black-white differences in sports performance appeared neither random nor accidental. It seemed to me that blacks performed relatively better in those sports which required "reactive" performance and whites performed relatively better in those sports that required self-paced performance. A young colleague, Allan Markle, joined me in testing that thesis. The results of these first studies were published in the *Journal of Personality and Social Psychology* (Worthy and Markle, 1970).

Reactive vs. Self-paced Performance

Reactive athletic performance involves making immediate and appropriate responses to changes in the situation confronting the athlete. Boxing, for instance, is a sport that places a high demand on reactive skills; the boxer, whatever his own plans, must be constantly responding to moves made by his opponent. The same thing is true, in varying degrees, for sports such as basketball, baseball and football. The player has little control over the timing of his response. He has little opportunity to "get ready" and must not unduly inhibit an immediate response.

Self-paced athletic performance, on the other hand, involves making responses in a situation that remains relatively static. Golf, for instance, is a sport that requires self-paced performance. Within broad limits the golfer can take his shot when he gets ready. An immediate response is not required and may be disadvantageous. The situation faced by the golfer changes from one shot to the next, but it does not change from one second to the next. The same thing could be said of bowling and other "nonreactive" sports. The terms "self-paced" and "nonreactive" will be used interchangeably.

Black/White Differences in Sports

Anyone who is familiar with sports is aware that blacks are very prominent in some sports, but not in others. In the United States sprinting and boxing are almost completely dominated by black athletes. In professional basketball more than half the players are black. To a lesser, but still impressive, degree blacks also excel as football and baseball players. No one doubts that this pre-eminence is fairly earned and based on superior performance. The explanation usually given for this pre-eminence is that blacks have traditionally had only a few career options open to them and one of the few was sports. Another explanation which has been advanced is that the hazards of the slavery experience weeded out all blacks except those who were most physically fit (see Kane, 1971). Yet another explanation advanced by psychoanalysts (Holloman, 1943) is that superior performance by black athletes results from sublimated hatred for the white man. According to this view sports serve as a safe way for the black man to defeat and punish the hated white man. More recently the effects of racial anatomical differences have been considered (Kane, 1971). Any or all of these explanations may have some validity, but I was struck by the additional consideration that the sports in which blacks excel seemed to be largely reactive ones.

There are few black golfers or bowlers. Golf and bowling are, of course, sports that are expensive to play and practice and blacks have often been denied opportunities to participate in these sports. Also, the lack of black heros in these sports may affect young blacks, who like all youth, pattern their goals and dreams on popular role models. Nonetheless, it seemed noteworthy that those sports in which blacks did not excel demanded self-paced performance.

We began, then, with the basic hypothesis that blacks, compared to whites, do relatively better on tasks requiring reactive performance than on tasks requiring self-paced performance. Markle and I could have tested the hypothesis by documenting the level of black and white participation in various sports, but that would not have gotten us far beyond our original casual observations and all the arguments about social exclusion. We decided that some of the social exclusion considerations could be handled by studying performance differences within the same sport. Professional baseball was chosen as one sport which would allow meaningful within-sport comparisons of black and white performance.

Baseball has a basic division of labor that can be viewed in terms of self-paced versus reactive performance. Pitching is self-paced and pitchers are evaluated almost entirely on their ability to pitch. Nonpitchers, on the other hand, are evaluated largely on their ability as hitters. Hitting, as well as other tasks done by nonpitchers, requires reactive skills. In general, the pitcher controls when each play begins; the hitter must react to the pitcher.

Consistent with our general thesis, we expected that, in terms of percentages, significantly more blacks could be classified as hitters than as pitchers. In order to test the hypothesis we obtained a list of all American-born players who were on major league rosters (excluding expansion teams) at the beginning of the 1969 baseball season. Of the 429 players 411 could be identified as to race and position played. Only 7% of the pitchers were black, but 24% of the nonpitchers were black. (This difference was statistically significant; $x^2=19.305$, $df=2$, $p<.001$). When we consider that blacks comprise about 11% of the American population, it is evident that blacks are underrepresented as pitchers, but have more than double their proportional representation as major league nonpitchers. Whatever the reason, the difference can hardly be attributed to chance.

This finding taken alone is still open to many interpretations. It is possible, for instance, that subtle biases operate to encourage blacks to be hitters and to discourage blacks from being pitchers. It is difficult to believe that any team today would knowingly discourage a good pitching prospect, whoever he might be, but there could be subtle influences that operate on the player even before he gets to the major leagues.

So the finding of this first study was suggestive of a racial difference in self-paced vs. reactive ability, but it was open to other interpretations. In order to choose between this and other interpretations it was necessary to gather more information. Two strategies were available. One was to study intensely all of the known factors that influence whether a player becomes a pitcher or a nonpitcher and attempt to control in our analyses for all factors that are unrelated to ability. That approach, studying a limited situation in depth, was not the strategy we chose to follow.

Instead, we used a strategy of research which is becoming increasingly popular in social science. The strategy, which is a "triangulation" approach, puts less emphasis on the purity of a particular study and more emphasis on the converging nature of evidence drawn from studies widely different in setting or methodology. A passage from the book, *Unobtrusive Measures* (Webb, Campbell, Schwartz and Sechrest, 1966) summarizes the rationale for such a strategy:

> The "outcropping" model from geology may be used more generally. Any given theory has innumerable implications and makes innumerable predictions which are unaccessible to available measures at any given time. The testing of the theory can only be done at the available outcroppings, those points where theoretical predictions and available instrumentation meet. Any one such outcropping is equivocal, and all types available should be checked. The more remote or independent such checks, the more confirmatory their agreement (p. 28).

So, rather than stay with baseball and try to identify and control all the variables that cause black players to become nonpitchers rather than pitchers, we tested our same general hypothesis about self-paced vs. reactive performance in another sport, basketball.

The overrepresentation of blacks as nonpitchers in professional baseball is exceeded by the overrepresentation of blacks in professional basketball. The percentage of black players in professional basketball is about five times the percentage of blacks in the population. It is interesting that this high percentage of black players is consistent with the fact that much of basketball is reactive in nature. Defensive basketball is almost entirely reactive and even on offense the player must respond to moves of his opponent and take shots in response to openings in the defense.

Basketball is largely a reactive sport, but not entirely so. One aspect of the game, shooting free throws, is self-paced. We expected, based on our thesis, that black players, compared to white players would do better at shooting field goals (partly reactive) than shooting free throws (self-paced). The data to test this were taken from information published by the National Basketball Association for the 1967-68 season. Only players who played a minimum of 1000 minutes during the season were included in the study. There were 53 black players and 45 white players who met that criterion. The percentages of field-goal accuracy and free-throw accuracy were computed for each player. In one statistical analysis, black and white players were matched on the basis of height so that that would not be a factor. The analysis indicated that blacks were slightly, but not significantly, more accurate than whites at shooting field goals (45% vs. 44%). Whites were significantly more accurate than blacks at

26

shooting free throws (75% vs. 71%, $t=2.499$, $df=36$, $p<.02$).

So although black players shot field goals as well as, or better than, white players they did not perform as well on the self-paced task of shooting free throws. This finding can not be explained in terms of exclusion since all players shot both field goals and free throws. There may be some other explanation but the self-paced vs. reactive thesis received support in findings from both baseball and basketball.

Our next study was a replication of the basketball study with college players. The sample consisted of 302 players whose coaches responded to a request for team statistics. Of the players, 229 were white and 73 were black. Each player answered a questionnaire about himself. The purpose of the questionnaire was to determine whether any differences in performance could be accounted for by differences in socioeconomic class or differences in father absence from the home. Since blacks and whites differ greatly on both those measures we wanted to know if such social factors could be the key to understanding the observed differences.

The analysis of differential performance on field goal shooting and free throw shooting yielded results very similar to what we had found for professional players. Again we first analyzed the data for all players and then reanalyzed the data for players matched on height. In both analyses the finding was that black and white players did not differ on field goal accuracy, but that whites were significantly more accurate than blacks at shooting free throws. The differences were not accounted for by socioeconomic class or father absence though there was a slight trend for black players from father-present homes to be more accurate at shooting free throws than were black players from father-absent homes.

Other racial studies have been done to test the self-paced vs. reactive hypothesis. Two studies (Bloomberg, 1972; Jones and Hochner 1973) have tested the hypothesis as it relates to hitting and pitching performance of major league baseball players. Their findings, together with ours, may be summarized as follows:

(1) Blacks are overrepresented as hitters and underrepresented as pitchers.

(2) Black pitchers, a select group in that their percentage among pitchers is lower than the black population percentage, perform as well as (Bloomberg) or better than (Jones and Hochner) the average white pitcher.

(3) Black hitters, though a non-select group in that their percentage among hitters is much higher than the black population

27

percentage, have much higher batting averages than white hitters.

(4) Taking all of the major league evidence together it is clear that blacks perform, relative to whites, better as hitters than as pitchers.

In another related study, Markle (1970) compared black and white college students in a laboratory setting on two golf putting tasks. One task, which involved hitting a moving ball, was designed to be reactive. The other task involved putting in the normal (self-paced) manner. On both tasks the ball was putted toward a target that indicated degrees of accuracy. The blacks, for some reason, did poorly on both tasks, but consistent with the hypothesis, they did relatively better on the reactive task than on the self-paced task.

Meanwhile, Dunn and Lupfer (1972) were completing a well conceived study at Memphis State University which tested the hypothesis under yet another condition. Their study differed from the ones we had conducted in that they used children and a sports activity that was new to all the participants. They designed a game especially for the study and checked the performance of black and white fourth-grade boys. The game involved a two-man arrangement in which one boy defended a soccer-like goal and the other boy tried to throw, kick, or roll a ball into the goal. If the ball went into the net, the boy on offense scored a point. If the defender could block, deflect, or otherwise keep the ball from going into the net, he scored a point. A player played half of each match on offense and half on defense. Each boy played matches against two opponents. His total points on offense and total points on defense were used as the measures of self-paced and reactive performance.

The results of this study were very clear; the black boys performed better on defense than they did on offense and they performed better on defense than did the white boys. The white boys performed better on offense than on defense and they performed better on offense than did the black boys.

After the games were completed, each boy was interviewed in order to obtain information about his athletic preferences and family background. The black and white boys did not differ in their preference for offensive vs. defensive play in the game they had just played. So, although there were large performance differences these were not related to differences in preference. When asked about other sports, there were some differences (e.g., the favorite sport of black boys was basketball; the favorite sport of white boys was baseball), but when asked about specific activities within sports that are self-paced or reactive, there were no racial differences in

28

preference for one type activity over the other.

Considering all the studies that had been done, the racial differences in self-paced and reactive sports were present to a degree over and above that which could be explained by any social factors that had been examined so far. We realized of course that there might be some powerful social determinants which had not been considered, but we also realized the possibility that the racial differences in self-paced vs. reactive ability could be genetic in origin. That possibility, being hard to test, has not been tested directly, but we have since collected some data which indirectly support the genetic interpretation.

Eye Darkness Differences
in Self-paced vs. Reactive Sports Performance

Just as our first series of studies were initiated to check a casual observation, the next series of studies also followed from a casual observation. I had observed that many quarterbacks in professional football appear to be light-eyed. Since there are few black quarterbacks and the job of quarterback seemed less reactive in nature than other positions in football, it occurred to me that athletic differences observed between blacks and whites might also be observed between dark-eyed and light-eyed Caucasians. The studies designed to test that notion were reported at a meeting of the Southeastern Psychological Association (Worthy, 1971).

The first test was simply to determine whether the percentage of blacks playing a particular position in football was related to the average eye darkness of whites playing that position. The hypothesis, of course, was that the greater the percentage of blacks playing a position, the darker would be the eyes of the whites playing that position.

Official photographs of players in the National Football League and American Football League (now merged into one league), were used to identify players as black or white and to rate eye darkness. Only players who had played five years or more were considered. The total sample included more than 300 players, but some of the pictures were not clear enough to allow a rating of eye darkness. Ratings were done "blind" (i.e., the rater did not know the hypothesis being tested) and made on a five point scale from very light (a score of 1) to very dark (a score of 5) (i.e., "very light", "light", "medium", "dark", "very dark"). The rater decided whether or not a picture was clear enough to rate.

29

The results of the first study are shown in Table 2-1.

TABLE 2-1

Percentage of Black Players and Rated Eye-Darkness of
White Players by Position in Professional Football (1969)

Position	Number of Players at each Position	Percent at each Position that are Black	Rank Order	Number of Whites that Could be Rated	Average Rated Eye Darkness of Whites	Rank Order
Defensive Back	(62)	58	1	(20)	4.25	1
Running Back	(38)	50	2	(12)	4.17	2
Receiver	(49)	33	3	(20)	4.10	4
Defensive Line	(59)	31	4	(28)	4.14	3
Offensive Line	(81)	15	5	(48)	3.79	5
Linebacker	(57)	7	6	(36)	3.64	6
Quarterback	(35)	0	7	(26)	3.31	7

The football positions in Table 2-1 are rank-ordered according to the number of blacks playing the position. The computed average eye-darkness of whites (whose pictures could be rated) in each position is also presented and ranked. The two sets of ranks are almost identical. The more blacks playing a position, the darker the eyes of the whites playing that position. (Rank-order correlation = .96, $p < .01$).

Since no football position is completely self-paced and since we did not attempt to rate them on the self-paced vs. reactive dimension, this finding could not be considered a direct test of the self-paced vs. reactive formulation. The rank order observed, however, with quarterback at one extreme and defensive back at the other, is not inconsistent with such a conception. One might also consider running speed as an important factor in this correlation since those positions with the more darkly pigmented players (blacks and dark-eyed whites) are also the ones that require the most speed.

In retrospect, it would have been better to have had several judges rate the pictures instead of just one. By using pooled ratings, the reliability of the ratings would have been slightly higher. After the various studies were completed, another rater did, in fact, rate the pictures so that we could get a measure of inter-judge reliability. The results indicated that ratings based on judgments of just one rater were adequately reliable (coefficients ranged from .75 to .91). An ideal situation would have been to have many judges rating multiple pictures of all players and all pictures taken under exactly comparable conditions. Better still would have been to have had an opthamological measurement of the amount of eye pigmentation. Such ideally precise conditions of eye darkness measurement were not obtained. However, it is easy to misunderstand the implications of that statement. The lack of precision in a rating instrument contributes error, but it is random rather than systematic error. The significance of the findings is not reduced by lack of precision in the eye darkness measure. If anything, the statistically significant relationships between eye darkness and sports performance that were found are more impressive since they were detected *even with a crude measure of eye darkness.*

The importance of this study was that it showed a parallel between race and eye darkness. That eye color varied with position played was an interesting finding, but the more important finding was that the pattern was similar to the one noted for racial differences.

The next question to be answered was whether eye darkness was related to other sports activities that were more clearly self-paced or reactive in nature. A comparison was made of average eye-darkness of a sample of white major league pitchers and white major league non-pitchers. There was no significant difference in eye-darkness between the two groups.

At that point it was decided to look not only at who played what baseball positions but also at how well they played the position. For these comparisons the ratings were combined to form a light-eyed group (ratings 1, 2, and 3) and a dark-eyed group (ratings 4 and 5). The middle rating was put in the light group because there were more dark than light ratings. This same manner of combination was used throughout all studies reported here. On batting average the dark-eyed and light-eyed nonpitchers did not differ. On pitching success there was a very significant difference between the two eye-darkness groups. The measure used was each pitcher's won-lost percentage. The average percentage of wins for the light-eyed pitchers was 52%. The corresponding average for the dark-eyed pitchers was 42% (t = 2.68, df = 52, p < .01). By the same token, 68% of the light-eyed pitchers had won as many or more games than they had lost. This was true for only 25% of the dark-eyed pitchers.

So, on this self-paced task, light-eyed clearly performed better than dark-eyed.

The next study involved professional basketball players. The same comparisons that were made earlier between black and white players were made between a sample of light-eyed and dark-eyed white players. The findings closely paralleled those of the earlier racial study. No difference was found on field goal accuracy, but on shooting free throws (the self-paced task) the light-eyed players were significantly more accurate than the dark-eyed players. The light-eyed players made 80% of their free throws; the dark-eyed players made 74% of their free throws (t = 2.83, df = 51, p < .01).

The final sport studied was professional bowling. Bowling appears to be almost entirely self-paced. The sample consisted of 5-year veterans of the Professional Bowlers Association tour for whom eye-darkness ratings could be obtained (N=55). The performance measure employed was the amount of prize money won the previous year. Light-eyed bowlers were found to have higher money winnings than dark-eyed bowlers. The average earnings were $19,900 for light-eyed bowlers and $11,400 for dark-eyed bowlers (t = 2.29, df = 53, p < .05). However, a subsequent study by Markle (1972), found no difference in performance between light-eyed and dark-eyed amateur bowlers in local leagues.

Considering all the studies of eye color and sports performance, significant differences were not found in all comparisons. However, every time a significant difference was found, it was in the same direction—dark eyes were associated with reactive performance and light eyes with nonreactive performance.

Supplementary Studies

In addition to the studies done to directly test the self-paced vs. reactive distinction we have done several post hoc analyses in an effort to throw light on what aspects of the situation are responsible for the differences noted. These were not formal studies and have not been replicated. They do, however, provide some tentative information to be considered.

For the professional football quarterbacks it appears that light-eyed quarterbacks have a significantly greater percentage of their pass completions go for touchdowns than do dark-eyed quarterbacks. Perhaps light-eyed quarterbacks, for some reason, are more effective passers than are dark-eyed quarterbacks when the team is in position to score. Or, perhaps, light-eyed quarterbacks, more than dark-eyed quarterbacks, wait for deep receivers to get clear.

A similar analysis of baseball suggests, first, that the pitching styles might be different for light-eyed and dark-eyed pitchers. Dark-eyed pitchers are often described as fast-ball pitchers. Among nonpitchers, dark-eyed tend to excel most often as outfielders; it is noteworthy that outfield play is very dependent on running speed.

We have also considered the possibility that success on self-paced athletic tasks depends to some degree on taking more time prior to executing the activity. Markle (1970) found on a golf putting task that accuracy was significantly correlated with time taken in putting. However, observations of time taken to bowl (Markle, 1972) and time taken to shoot free throws in basketball (Sheedy, 1971) indicated no relationship between time taken and accuracy of performance.

To summarize, a number of studies of sports performance have provided evidence to indicate that dark-eyed people perform relatively better on reactive athletic tasks and light-eyed people perform relatively better on self-paced athletic tasks. These differences refer to group averages, of course, and not to individuals as such. There are many individual exceptions to the group averages.

We believed at this point that the self-paced vs. reactive distinction was an important one and was probably related to genetic differences in ability. Why these differences should be related to eye darkness, however, was an intriguing mystery, intriguing enough in fact, to keep us searching for other correlates of eye darkness that might increase our understanding of the total picture.

References

Bloomberg, M. Achievement differences between black and white professional baseball players in 1970. *Perceptual and Motor Skills,* 1972, *34,* 269-270.

Dunn, J. R. and Lupfer, M. A comparison of black and white boys' performances on self-paced and reactive sports activities. Paper read at Southeastern Psychological Association, Atlanta, April, 1972.

Holloman, L. L. On the supremacy of the Negro athlete in white athletic competition, *Psychoanalytic Review,* 1943, *30,* 157-162.

Jones, J. M. and Hochner, A. R. Racial differences in sports activities: A look at the self-paced versus reactive hypothesis. *Journal of Personality and Social Psychology,* 1973 (in press).

Kane, M. An assessment of 'black is best.' *Sports Illustrated,* 1971, *34,* (3), 72-83.

Markle, A. Self-paced and reactive performance as a function of activation level and competition. Unpublished master's thesis, Georgia State University, 1970.

Markle, A. Effects of eye color and temporal limitations on self-paced and reactive behavior. Unpublished doctoral dissertation, Georgia State University, 1972.

Sheedy, A. The optimal limits of concentration time relative to success in basketball free-shooting during international competition. *International Journal of Sport Psychology,* 1971, *2,* 21-32.

Webb, E. J., Campbell, D. T., Schwartz, R. D. and Sechrest, L. *Unobtrusive measures: Nonreactive research in the social sciences.* Chicago: Rand McNally and Company, 1966.

Worthy, M. Eye-darkness, race and self-paced athletic performance. Paper read at Southeastern Psychological Association, Miami, April, 1971.

Worthy, M. and Markle, A. Racial differences in self-paced versus reactive sports activities. *Journal of Personality and Social Psychology,* 1970, *16,* 439-443.

Chapter 3

Animal Eye Color and Behavioral Reactivity in Birds of Prey

After finding that eye-darkness was related to sports performance we began to wonder about other areas of performance. Since we were not sure what areas might prove most fruitful, we began checking out different areas as they occurred to us. Another colleague at Georgia State University, Dr. James Dabbs, worked with me in obtaining eye color and performance data of men in naval flight training. The Naval Flight Training Center at Pensacola, Florida made their records, which included eye color, available to us for study. We originally thought that eye color might be related to performance evaluations routinely made of all students. Eye color was not, however, related to any of the available measures of flight performance.

Another study that was a "shot in the dark" also missed. Since eye darkness measures could be obtained from pictures and the bigger the picture the easier the rating, we were prompted to study eye darkness and voting behavior of U. S. Congressmen. One does not find a readier source of 8 X 10 glossy mug shots than among members of Congress. I wrote to all members of Congress and asked for a photograph. Almost all of them promptly complied with my request. I had, in the past, used representative samples; this was the first time I used a sample of representatives. Eye darkness was rated for each Congressman and correlated with ratings of conservative versus liberal voting records in Congress. There was no relationship whatever.

A certain amount of floundering is a necessary part of doing research in new areas that have not been charted with earlier empirical and theoretical guideposts. One can only keep looking for leads and hope to pick up the trail again.

As it turned out, getting back to a fruitful line of research involved breaking the mental "set" of thinking only about human

35

performance. Since animals also differ on eye color it seemed reasonable to take a look at animal behavior as it related to eye color. If eye color differences were genetically linked to behavioral differences the phenomenon might be observable in animals. Such a parallel, between animals and humans, if it were found, would tend to indicate a basic natural difference rather than a superficial, chance, or cultural difference. In addition, animal studies, it seemed, might provide some clues as to why or how natural selection had operated to select light eyes in some cases and dark eyes in others.

I began to look for books on animals that might provide data for analysis. What was needed were sources that gave information on a number of closely related species differing in eye color. It was essential to have information on eye color for each species and information about the habits of each. My hope was to find differences in habits that might be relevant to our earlier findings on self-paced and reactive activity.

Hunting Tactics of Eagles, Hawks and Falcons

Luckily, I found a source that was very good for such studies. In 1968, the National Audubon Society had sponsored the publication of a two-volume work, *Eagles, Hawks and Falcons of the World*, written by Leslie Brown and Dean Amadon. This book provides a species-by-species description of birds of prey and their habits. Iris color was routinely included as a part of the description and one of the categories, "food", provided information about the kind of prey eaten by that species. I had a vague notion that different types of prey would require different hunting skills and that these skills might also be classified as self-paced or reactive. Based on the studies in human sports performance I anticipated that the dark-eyed birds would exhibit the more reactive hunting skills.

Fortunately, there is a wide distribution of eye colors among birds of prey. They range, in general, from pale yellow to dark brown. It took some time to settle on a quantitative scale of animal eye darkness based on eye color, but in time a five point scale was chosen and used in all studies of animals. The scale was as follows: 1 = blue, green, yellow, white, pink, amber, 2 = orange, red, hazel, 3 = reddish brown or light brown, 4 = brown, 5 = dark brown or black. If members of the species varied in eye color, the lightest color was used for the rating. If the eye color for males and females differed, we used the male eye color. The scale reflects degree of iris pigmentation. The higher the rating the greater the amount of pigmentation.

Eagles, hawks, and falcons are successful predators that take a variety of food. Some species specialize in just one type of prey and other species feed on a variety of foods. With almost all of the species it was possible to determine a "chief food" (e.g., birds, snakes, etc.). Figure 3-1 shows the relationship between eye-darkness ratings and categories of food taken as chief food by ten or more species.

Taking the foods in order, we see first that not many species specialize in catching fish but that those that do are mostly light eyed. Catching fish requires of the bird patience and inhibition of premature attack. This point was made clear by Brasher (1968) in his description of Osprey hunting characteristics:

Old birds seldom miss a strike, but the youngsters often fail to capture their quarry. One reason for this is that the experienced bird will let many chances go by that he does not consider sure enough, while the younger bird takes a start at any fin which shows (p. 33).

There is also a definite tendency for reptiles to be taken as chief food by light-eyed birds. This prey also must be hunted patiently and taken by surprise. If a snake or lizard is attacked prematurely it can, given a moment's warning, easily escape into places that cannot be reached by the bird. Birds of prey that were overly reactive would not be successful as hunters of reptiles. It is consistent with our ideas about eye color and reactivity in general that reptile eating and light eyes have evolved together in birds of prey.

The same skills are important in catching mammals, but perhaps not to the same degree. Mammals also are hunted by means of surprise tactics, but the margin of acceptable error is probably somewhat greater. That is, mammals often can be caught even if they get a slight advance warning because they do not have a ready place of escape. However, the bird that is patient can often increase its chance of success. For example, a hawk will often sit watching a mouse at a hole in the ground. As long as the mouse stays near the hole, the hawk will not move. However, when the mouse finally ventures several feet from the hole, the hawk attacks and catches the mouse before it can get back to its hole. The successful hawk must wait until the moment is right. The eye color distribution in figure 3-1 for those species that prey on mammals shows that again more light-eyed than dark-eyed species feed on this food, but the contrast is not so great as it was for fish and reptiles. The curve is very close to the curve of eye color distribution for all eagles, hawks, and

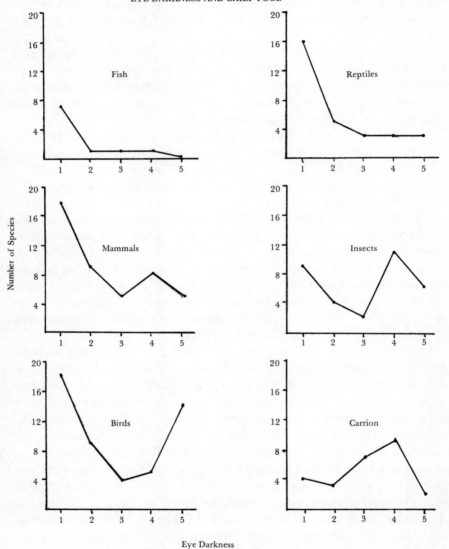

Figure 3-1

EAGLES, HAWKS AND FALCONS:
EYE-DARKNESS AND CHIEF FOOD

falcons. For, although most birds are dark-eyed, most birds of prey are light-eyed. Mammals represent a prey that may be taken by birds with light eyes or dark eyes, but the advantage (based on number of species) goes to the light-eyed.

The three types of prey, then, that are most often caught by means of patient waiting and surprise are caught primarily by light-eyed species. With the next type food to be considered, insects, surprise and waiting are not so important. Insects are typically taken immediately with no delay and reactive skills are clearly an advantage in catching them. In Figure 3-1, the curve for insects reveals a clear shift from the curves we have already considered. The percentage of dark-eyed species that take insects as chief food is much higher than is the case for fish, reptiles, or mammals, which is consistent with the eye color reactivity hypothesis.

Some light-eyed species also eat insects. What is interesting is that the light-eyed and dark-eyed species differ in how they catch insects. Two-thirds of the light-eyed species catch some or all of their insect food on the ground. This is true for fewer than one-third of the dark-eyed species. The dark-eyed birds take insects "on the wing," which certainly requires more speed and reactive skill than does catching insects on the ground. This relationship between ground vs. "on the wing" feeding and eye darkness is statistically significant (x^2 = 4.99; $df = 1$; $p < .05$).

Many species of eagles, hawks and falcons feed on other birds. As can be seen in the lower left-hand chart of Figure 3-1, success in catching birds depends on being either very light-eyed or very dark-eyed. Two very different techniques are used to catch birds. One depends on direct and bold attack and the other depends on surprise. Dark-eyed species are much more likely to use direct attack and light-eyed species are much more likely to use surprise. Among the birds of prey that chiefly feed on other birds, most species fall either in the *Falco* or the *Accipiter* genus.

Those species in the *Falco* genus tend to have brown or dark brown eyes and catch birds "on the wing". In pursuit they get above the prey and then swoop down ("stoop") on it with great speed. If they miss, they gain altitude and strike again. They are graceful fliers that have been prized for centuries by those who engage in the ancient practice of falconry. The best known and perhaps most admired species in the *Falco* genus is the Peregrine Falcon:

> The perfect skill of this species in flight is always a joy to
> watch. A Peregrine is hardly ever seen off balance or taken
> aback, even in violent wind, while they are capable of hair

breadth accuracy at great speed when stooping either in play or to kill. . . . The speed of their stoop has been variously estimated at 150-200 miles an hour by observers, but recent evidence obtained by more accurate methods indicate that they may achieve 275 miles per hour, and that they are capable of breathing even at this terrific speed. (Brown and Amadon, 1968, p. 853).

Not all species in the *Falco* genus are dark-eyed; a few are very light-eyed. Those species do not depend on attacking birds for food but subsist mainly on insects taken on the ground. It is only the dark-eyed falcons that prey on birds in the air.

Among the hawks of the *Accipiter* genus, most of which are light-eyed, there are many species that also catch birds. However, their method of hunting differs markedly from that of the bold and reactive falcons. Perhaps because they are slower, the *Accipiters* depend more on surprise. They are mostly woodland birds that ambush their prey from cover. These light-eyed hawks use a variety of indirect methods to achieve surprise. For instance, Brown and Amadon (1968) report that the European Sparrow Hawk, when it spots a group of sparrows feeding on the ground, approaches, "preferably using a hedge or bush to provide cover for an attack, approaching low down behind such cover, and suddenly flicking over the bush at the last moment and taking its prey by surprise" (p. 477).

As mentioned before, members of the *Falco* genus that differed from the group norm in eye darkness also differed in diet and hunting habits. A similar exception is noted for the *Accipiter* genus. The Grey Frog Hawk differs from most species in the *Accipiter* genus by being darker eyed. It also differs in hunting method, as noted by Brown and Amadon (1968), "As it has the peculiarity of feeding mainly on frogs it hunts chiefly in open ground, either in swamp or rice fields and does not have to resort to the swift rushes and stealthy surprise tactics of most sparrow hawks" (p. 515).

So dark-eyed hawks use bold attack and light-eyed hawks use stealth, and surprise, but what of those that are intermediate in eye darkness? How do they catch prey? The answer is that they often do not. Hawks that are toward the midrange of eye darkness (reddish brown on our scale) are the species that are scavengers, subsisting mainly on carrion. Note that the last curve in Figure 3-1, that of carrion, shows many species in the midrange of eye color. That curve is very unlike the others and is the only one that approaches a symmetrical distribution. This characteristic of the distribution is even more pronounced, as seen in Figure 3-2, when one plots the

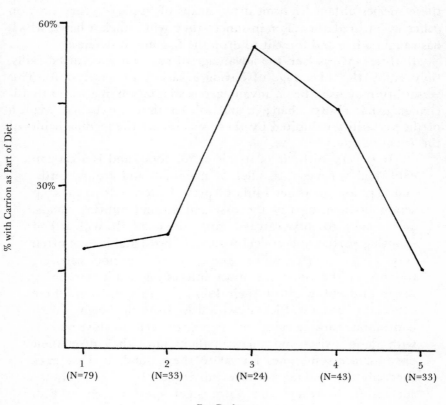

Figure 3-2
PERCENTAGE OF SPECIES OF EAGLES, HAWKS
AND FALCONS AT EACH DYE DARKNESS RATING
WITH CARRION AS PART OF DIET

percentage of species at each eye color that takes carrion as any part of the diet. Those species in the midrange of eye color are about three times as likely to eat carrion as are those species with very light or very dark eyes.

Apparently, those species with eye color in the midrange are lacking in the speed and agility to be reactive hunters and in the inhibition and patience to stalk or ambush prey. Lacking a specialized skill, they survive by being opportunistic and omnivorous. In addition to carrion, foods eaten more by these species also include eggs and nestlings, snails and crabs, etc. All of these foods are ones that do not require a high level of hunting skill. Also, a number of these species obtain fresh meat by means of "piracy"; they prey on other avian predators. For instance, they will attack a hawk which has caught a fish and force it to drop the fish in self defense.

If these species lack the advantage of superior specialized skills, they enjoy the advantage of having a variety of lesser skills. This versatility may itself be an advantageous adaptation in a world that is changing, has always changed, and will continue to change. Typical of the versatility exhibited by these species are the feeding habits of the Tawny Eagle:

It is very catholic in its tastes for food, and is a carrion eater and scavenger, a killer of mammals and ground birds, and a pirate on other birds of prey. It descends to pick up scraps in townships in the east and around hunting camps, and comes to any carcass, large or small. It will attend shooting parties and collect wounded birds, and is then often very bold. ... (It) walks about on the ground catching rodents in the open. Its most interesting habit is robbing other birds of prey of their kills .... In pursuit of these others it usually flies downwards from a height, with continuous barking calls, and thereafter keeps in close pursuit with rapid twists and turns in flight until they drop their prey, or are sometimes forced to the ground; it sometimes succeeds in making vultures disgorge. It is not always successful, however, and with larger species such as Fish Eagles must often wait till these have finished feeding before eating the remains. It is not invariably the sluggish and scavenging sort of bird it is sometimes made out to be, and can on occasion kill active mammals, lizards, snakes, and even birds; it has been known to attack successfully migrating flight of flamingos in daylight, striking them on the wing with a spectacular stoop. It probably, however, gets most of

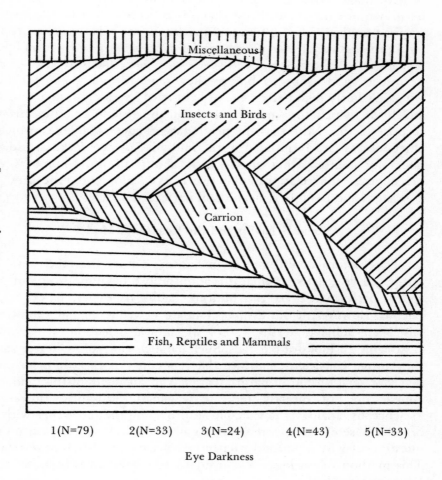

Figure 3-3
PERCENTAGE OF SPECIES OF EAGLES, HAWKS
AND FALCONS AT EACH EYE COLOR BY VARIOUS
CATEGORIES OF CHIEF FOOD

% of Species Taking Each Food

Miscellaneous

Insects and Birds

Carrion

Fish, Reptiles and Mammals

1(N=79)   2(N=33)   3(N=24)   4(N=43)   5(N=33)

Eye Darkness

its prey by piracy or scavenging (Brown and Amadon, 1968, p. 654).

Considering again the total picture of chief food for various species of birds of prey there is another way that we can look at the data. In Figure 3-1 we started with the food and noted the number of species at each eye darkness rating that ate more of that food than any other. In Figure 3-3 we have started with the species at each eye darkness rating and noted the *percentage* of those species that took various chief foods. Both ways of viewing the data, in absolute terms or percentage terms, are legitimate and in this case the conclusions to be drawn are the same. Note that foods which must be caught by surprise tactics, fish, reptiles and mammals, are taken as chief food by a progressively lower percentage of species as one goes from light to dark eyes. The darkest-eyed species are more likely to take birds or insects, and as noted before they tend to take them "on the wing." A fairly high percentage of light-eyed species also take insects and birds, which they mostly catch on the ground or by means of surprise tactics. Note that the percentage of species taking carrion as chief food is clearly inflated in the midrange of eye darkness. So, it does not matter whether we talk about absolute numbers or percentages, the relationship between eye color and type prey is consistent and fits into a meaningful pattern.

Animal Behavior and the Reactivity Model

The pattern of differences in hunting tactics fits nicely into a self-paced vs. reactive model. It is admittedly a big jump from human sports performance to the hunting habits of eagles, hawks, and falcons. However, if one assumes human, as well as animal, evolution has been subject to the forces of natural selection, it is only reasonable to look for parallels. Such parallels can, to the degree that they prove consistent, indicate the generality of the eye color and behavior relationship in nature. Parallels or analogies of this kind, of course, do not directly indicate a genetic causality. However, "science by analogy" is valuable in providing the researcher with a broader base of facts from which to make theoretical forays into the question of why a particular pattern is observed in different settings. This method of analogy or converging triangulation is basically the Darwinian method.

The way in which Darwin sought to prove his theories bears little resemblance to what is often taught as standard scientific method. No crucial experiments, no statistical test, no quantitative predictions. Instead his method was to establish the probable truth of a proposition by means of converging, independent lines of evidence, some from his own experiments (carried out to test the hypothesis), some from the observations of others: "The line of argument often pursued throughout my theory is to establish a point as a probability by induction and to apply it as hypotheses to other points and see whether it will solve them." The role of experiment was not—as in the proto-typical 'crucial' experiment—to provide the keystone to an arch of deductive reasoning, but simply to generate additional evidence, to find some more pieces of the jigsaw puzzle (Staddon, 1971, p. 690).

Keeping in mind the triangulation method of analysis it is interesting that the tactics of light-eyed hawks are similar in nature to the tactics of another group of light-eyed animals, cats. Cats use a variety of methods to surprise their prey. Some use cover to stalk the prey and others drop on the prey from rocks or tree limbs. One species of jungle cat has a unique method of catching monkeys. The big cat droops across a tree limb and plays dead. The monkeys, being curious, approach closer and closer to view the dead cat. The monkey with the greatest amount of curiosity gets to be the cat's supper.

Based on the converging nature of the findings, I reasoned that eye color may be related to the method of survival employed by prey and predator. Two contrasting strategies may be thought of as react-approach-flee and wait-freeze-stalk. The reacters survive by means of an appropriate active response. The nonreacters survive by delaying or holding back an active response. A nonreactive prey may freeze until the predator is gone, and likewise a nonreactive predator may stalk, which is basically a strategy of freeze-approach-freeze.

Which of these strategies is employed may have been determined by the type of environments in which the animal's ancestors lived. If delaying had greater survival value than reaction, the group would have evolved into a population of nonreacters. If quick responding led to survival then the population would have become one of reactors.

It appears that the forces in nature that select a react-approach-flee tendency also select dark eyes and that the forces in nature that select the wait-freeze-stalk tendency also select light eyes. The next two chapters will present additional examples of this relationship as observed among animals.

45

References

Brasher, R. *Birds and trees of North America.* New York: Columbia University Press, 1968.

Brown, L. and Amadon, D. *Eagles, hawks and falcons of the world.* New York: McGraw-Hill Book Company, 1968.

Staddon, J. E. R. Darwin explained: an object lesson in theory construction, Review of *The triumph of the Darwinian method* by M. T. Ghiselin. *Contemporary Psychology* 1971, *16,* 689-691.

Chapter 4

Eye Color and Reactivity in Other Birds

Eagles, hawks, and falcons were studied in detail because of the excellence of the available source book. No other group of birds was studied in that detail, but other sources of information were read to determine whether the conclusions based on birds of prey were consistent with observations of other birds. The main source used was the work by Pearson (1936), *Birds of America.* This included information on more than 450 species for whom eye color was provided. The species were grouped by families of birds. No effort was made to make a complete quantitative analysis, but the descriptive material for each family of birds was studied with the eye color/reactivity relationship in mind. The results of that study are very interesting.

The approach here was to look for feeding habits that represented the extremes of react-approach-flee and wait-freeze-stalk. If the observations based on eagles, hawks, and falcons were representative of other birds, then one might expect to find that birds with the more reactive habits would be dark-eyed and birds with the more nonreactive habits would be light-eyed.

Feeding Habits

The most reactive method of getting food that was noted was catching insects on the wing. The species that get their food in that way are virtually all dark-eyed. This includes such groups as Swifts, Flycatchers, and Swallows. Those who observe these birds are struck by their dash, grace, and what appears, in human terms, to be confident exuberance. Brasher's (1968) description of the Cliff Swallow includes this observation, "They then ceased from labour for a few hours, amused themselves by performing aerial evolutions, courted and caressed their mates with much affection, and snapped at flies and other insects on the "wing" (p. 93). Light-eyed insect-eaters, on the other hand, are less dashing and more likely to forage on the ground as do the Thrashers and Towhees or obtain insects from the bark of trees as do the Woodpeckers. The observation that birds that feed "on the wing" are dark-eyed illustrates once again the connection between dark eyes and the react-approach-flee adaptation.

Turning to the other extreme, there is one family of American birds, the Herons, that clearly illustrates the wait-freeze-stalk adaptation. The Heron family includes Bitterns and Egrets. There are eleven species in the United States and all live near water. All are light-eyed; eight species were rated "1" on the 5-point scale and 3 species were rated "2". Gladden (1936) gives a description of the hunting habits of the Great Blue Heron which depicts well the wait-freeze-stalk approach:

> Much of this fishing the Heron does without stirring from the position he takes in shallow water among reeds or near the shore. Motionless as a statue he stands, his long neck doubled into a flattened S and his keen eyes searching the water nearby. As a frog or fish approaches he holds his rigid position until the creature comes within striking range, and the Heron knows what that is to a small fraction of an inch. Then suddenly the curved neck straightens out and simultaneously the long, rapier-like bill shoots downward with a stroke which is quicker than the eye can follow and seldom misses its mark. In a second the fish or frog has disappeared, and the fisherman has resumed his statuesque pose. Again, the great bird may be seen stalking slowly through shallow water, lifting each foot above the surface, and sliding it into the water again so gently as to cause hardly a ripple; and woe to the crawfish or salamander that does not observe that approach (p. 185).

This use of patient waiting and stealth is employed by Herons mostly in pursuit of fish and other water creatures. Occasionally, however, they go on land and stalk rodents.

As mentioned before, three of the species are slightly darker-eyed than the other species in the family. These three include two species of Night Herons and the Louisiana Heron. Pearson (1936) says of the Night Heron:

> Its hunting methods differ from those of its relative, the Great Blue Heron. Instead of standing rigid, and knee-deep in water, as that big fisherman does, the Night Heron moves about briskly, holding its head lowered and its neck curved,

48

all ready for the quick stroke which means death to the frog or fish at which it is aimed (pp. 194-195).

Brasher (1968) made a similar observation, "They (Night Herons) lack the patience of other herons, and instead of waiting for prey to come within range, pursue their victims with rapid steps" (p. 14).

If we think of those species of birds that feed entirely on creatures in flight as representing the react-approach-flee end of the continuum and those like the Herons, that depend entirely on stalking, as at the other end of the continuum, that leaves the vast majority of species falling somewhere between those extremes. Most species of birds are near the reactive end of the continuum. They are brown-eyed and eat a mixed diet of seeds and insects. Those birds more clearly in the midrange (reddish brown) of eye darkness, similar to the situation noted with birds of prey, have evolved feeding habits that do not depend very much on either stalking or reactive skills. The Raven feeds on carrion and garbage; the Turkey eats bird eggs—even its own at times; the Red Headed Woodpecker eats some nestlings and eggs. The Common Tern flies over the ocean feeding on the small fry driven to the surface by schools of large fish. The Northern Pholarpe whirls around in shallow water and then feeds on the tiny marine life that has been stirred up. The most common adaptation made by birds with medium eye darkness, however, is to feed on the various foods, such as grains and fruits, that are more available because of man. This food supply also includes, of course, the increased population of insects that is drawn to cultivated areas. A most interesting adaptation is that of the Cuckoo, which solves part of the food problem by laying its eggs in the nest of other birds and letting them feed and rear the young.

Escape Tactics

Just as the light-eyed and dark-eyed birds differ in their feeding habits, they also seem to differ in their manner of escape from predators. The dark-eyed birds, when they are alarmed, flee from possible danger. Since most birds are relatively dark-eyed, flight is the common technique of escape.

When we look for escape techniques that deviate most from the common practice of reactive flight, we come again to the Heron family. Some of the Heron species are among the best practitioners of the freeze method of escape. One member of the family, the Bittern, is perhaps the best adapted of any bird for this maneuver. When the Bittern, which lives in marshy land, becomes alarmed, it

49

points its long bill straight up to the sky and becomes perfectly motionless. Successful concealment results from the fact that the color pattern of feathers is such that with bill pointed skyward the bird blends perfectly into the grass and reeds of the environment. Before leaving the freeze position the Bittern surveys the scene by turning its head to look all around. These moves are made so slowly that they have been likened to the movement of the minute hand on a clock.

There are other birds that also use freezing or inhibitory defenses to some degree. Most of these are in the light to middle range of eye-darkness.

Many birds in the midrange of eye-darkness, however, are deficient in defensiveness. When in danger, they seem not to have the instinct to freeze *or* to flee. Those species that are hunted by man are very vulnerable. The Labarador Duck, for instance, was so tame that hunters had no trouble shooting it and it is now extinct. Also extinct in the United States, or nearly so, is Mearn's Quail, a bird that was known as the "Fool Hen" because it would allow a man to walk up and knock it in the head with a stick. The Artic Three-Toed Woodpecker also fails to hide or flee when approached by hunters. The Russet-Backed Thrush and the Wood Pewee allow unusually close approach to the nest by humans and the nesting Blue Headed Vireo has been known to allow a person to approach and stroke its back as it sat on the nest. All of these species have (or had) reddish or light brown eyes.

After concluding that birds in the midrange of eye-darkness are often weak in defensive responses, I recorded eye color of species listed in *Extinct and Vanishing Birds of the World* (Greenway, 1967). If the idea about eye color and lack of defensiveness was correct then one might expect many midrange eye colors among those species that survived until recently, but have become extinct since man and his animals became a significant and dangerous ecological force. Descriptive information on extinct species is often limited, but Greenway does list eye color for 18 extinct species. Whereas only about 30% of the bird species in the world have eye colors darker than yellow but lighter than brown 17 of those 18 extinct species are in that midrange. This phenomenon warrants further study by those interested in ecology and the preservation of wildlife.

Some more insight into the behavior of birds with eyes intermediate in darkness is provided by a short study I did of land birds that live in India. The book *Whistler's Popular Handbook of Indian Birds* by Kinnear (1949) was used as a source. The term

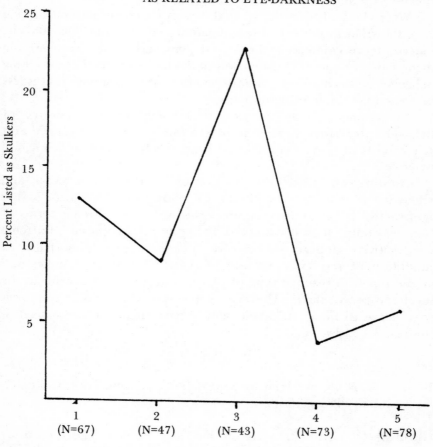

Figure 4-1
SKULKING BEHAVIOR IN SPECIES OF INDIAN BIRDS
AS RELATED TO EYE-DARKNESS

51

"skulking" is commonly employed in that book to describe birds that do not flee when approached, but move about in a stealthy or slinking manner. This study involved a simple tally of the number of light-eyed and dark-eyed species identified as "skulkers". It was anticipated that more light-eyed than dark-eyed species would be described as skulkers and that was indeed the case. Of the light-eyed species 15% were so described as compared to 5% of the dark-eyed species (N = 308 species, X^2 = 7.436, p < .01).

When the analysis was broken down into each eye-darkness rating as is shown in Figure 4-1, it was apparent that skulking behavior was, indeed, more common in light-eyed than dark-eyed birds, but the highest percentage of skulkers was in the midrange. Perhaps skulking behavior as it is used here indicates a nonreactive response that is less extreme than the freezing response.

To summarize, the relationship of eye color to degree of behavioral reactivity appears to be the same for birds in general as it is for birds of prey. Eye darkness is positively related to behavioral reactivity.

Another aspect of the relationship between eye color and reactivity which is made plain by the study of bird behavior is that nonreactive behavior may at times involve more extreme forms of non-responding than is connoted by the term "self-paced". Extreme nonreactivity appears to involve a complete holding back or inhibition of response. An athlete that "freezes" or "chokes up" would not be a reactive type of player, but he would not make the ideal self-paced player. However, some ability to inhibit responses may prove to be a critical element in performance of self-paced or nonreactive tasks.

References

Brasher, R. *Birds and trees of North America.* New York: Columbia University Press, 1968.

Gladden, G. Great Blue Heron in T. G. Pearson (Ed.) *Birds of America.* Garden City, New York: Garden City Publishing Company, 1936, Vol. 1, 184-185.

Kinnear, N. B. *Whistler's popular handbook of Indian birds.* (4th ed.) Edinburgh: Oliver and Boyd, 1949.

Pearson, T. G. (Ed.) *Birds of America.* Garden City, New York: Garden City Publishing Company, 1936.

Chapter 5

Eye Color and Reactivity in Nonhuman Mammals

The behavior of mammals in regard to eye color and reactivity was found to be consistent with that described for birds. It was not as easy to compare species of mammals as it was to compare birds. Whereas iris color is usually provided in the description of a bird species, it is rarely provided in the description of mammal species. There is some information about wild animals, differences in breeds of domesticated animals and differences in behavior of different strains of laboratory animals that proved helpful.

Some Examples From Nature

It is general knowledge that cats, both wild and domestic, are light eyed. Most have blue, green or yellow eyes. Cats also represent the epitome of the wait-freeze-stalk description. They are very patient and stalk slowly and quietly. Even domestic cats, which have not been subject to the pressures of natural selection for many generations, retain the ability and inclination to stalk any prey that is available.

The cat is the best known practitioner of the indirect (non-reactive) method of hunting, but wolves, coyotes, and foxes, which also tend to be light-eyed, also use indirect, waiting, nonreactive methods of attack and escape. These habits have earned the fox a reputation for slyness. Of the wolf, Asdell (1966) says, "Wolves are usually cowardly; that is, they are cautious and tend to be fearful of unfamiliar objects" (p. 169). Asdell also reports that wolf-dog hybrids differ from most dogs in a tendency to more readily respond to stress with a "passive defense reaction."

Slyness, patience, indirectness, and passivity are traits that are used to advantage by the wolf as it hunts game. It does not approach the prey directly, but uses circling, zig zag, and ambush tactics. These tactics may be employed to take domestic animals as well as wild

53

game.

Arctic explorers have many times recorded instances of wolves entering Eskimo villages and carrying away dogs. There also have been accounts of wolves enticing dogs away from the protection of their owners by manifestations of friendship and antics of play, only to turn on them and tear them to pieces when they had succeeded in luring them beyond the reach of their masters. Sled dogs have been known to succumb to the enticements of a female wolf in season, only to be devoured by members of the pack lying in wait (Davis, 1956, p. 7).

In contrast to cats and wolves, most of their prey are dark-eyed and depend on quick reactions to avoid capture. For some of these animals the reactive response does not become dominant until the animal reaches a certain level of maturation. The young of goats, antelopes, and deer employ the "freeze" technique to escape danger. I don't know that this change in response style is related to change in eye pigmentation, but it is true, in the deer at least, that during maturation lightly pigmented spots in the coat darken and disappear. It is also a fact, at least in some animals, that spotting of the coat is related genetically to decreased pigmentation of the eye.

When one thinks of the freeze reaction in adult animals, he thinks first of the opossum. The opossum is noted for an extreme freeze response which resembles death. The opossum, which is a primitive form of nocturnal mammal, is one of the least pigmented of all animals; many opossums are albinos. In addition, the eye of the opossum has no dilator or iris which is the part of the outer eye normally pigmented. Consequently the pupil is always wide open, which makes it very sensitive to light which is similar to the condition observed in animals with lightly pigmented irides.

Behavioral Reactivity in Dogs

The examples, based on mammals, which have been given so far may be suggestive of a pigment-reactivity relationship, but as evidence, they necessarily lack the rigor of a quantitative analysis. With one category of mammals, dogs, it was possible to provide a more rigorous analysis. Several handbooks are available that provide information, including eye color, on different breeds of dogs. The eye color is sometimes not given or given in an unusable form such as desired color rather than actual color. Moreover, eye color is highly variable in some breeds. Even with these limitations it was still possible to obtain eye color information on enough different breeds to make meaningful comparisons. The handbook used was *Dogs of the World* by Sneider-Leyer (1970) from which eye color information was available on 204 different breeds. For those breeds in which eye color was variable the lightest color listed was used as the standard. This was the same procedure that was used in rating the eye color of birds. Likewise, the same rating scheme employed with birds was used to assign eye-darkness ratings to different eye colors. When the ratings were combined the "light-eyed" category included all colors light brown or lighter; the "dark-eyed" category included everything listed as dark, brown, dark brown, or black.

Given that breeds of dogs differ in eye colors that can be categorized as light and dark, the question becomes whether there are also differences in behavior. With dogs we are dealing with various subspecies (breeds) that have been bred for desired behavioral traits. Depending on the needs of the people involved, dogs were bred for use as pets, watchdogs, herders, hunters, etc. Only those dogs that exhibited the desired behaviors were used for breeding purposes. In this way different breeds were developed that vary greatly one from another. However, breeds in different countries that were bred for the same purpose tended to resemble each other not only behaviorly but also physically. For instance, coursers from different countries, selected for superior speed, tended to have the lean greyhound appearance. More recently with the advent of dog shows and the decline in the use of working dogs, selection for breeding is more often made directly on the basis of appearance than was the case previously. So, even though physical characteristics in different breeds may today be purposefully maintained these different characteristics evolved originally because they were more common in the dogs that had the desired behavioral characteristics.

With that in mind, it appears that dark eyes were more likely than light eyes to be associated with traits desired by man. This assumption is based on the fact that most breeds of dogs, especially those valued as pets or watch dogs, are dark eyed.

One might speculate about the degree to which dogs in general were selected for their reactive abilities, but it is possible to make a more convincing argument by looking at the one category of dogs that includes dogs clearly selected for their nonreactive (wait-freeze-stalk) abilities. If breeds in that category have eyes lighter than those of other breeds of dogs, the eye color-reactivity hypothesis is supported. The category of dog to which I refer is that of "Pointers and Setters".

Pointers and setters were selectively bred for their ability to locate game and then "freeze". By "freezing" the dog did not frighten away the game, usually birds or rabbits, and the posture of the dog indicated to the hunter the location of the game. The hunter could then bring up swift coursing dogs to chase the game or he could get into position to shoot before the game was flushed. Pointers and setters are still used for that purpose. Pointers indicate the presence of game by "pointing" in an upright, rigid position. Setters, on the other hand, freeze in a crouching position similar to the stalking position of a cat. Good pointers and setters are reported to be able to hold the freeze position for more than an hour without moving.

Compared with other breeds we expected that pointers and setters would be lighter eyed. Among the 204 breeds in *Dogs of the World* that could be rated as to eye color, 33 breeds were in the "Pointers and Setters" category. The eye color ratings were combined into a light-eyed category and dark-eyed category as described previously. The comparison of pointers and setters with all other breeds is shown in Figure 5-1. Whereas only 23% of other breeds were light-eyed, 73% of the pointers and setters were light-eyed. Statistically, this difference was highly significant ($x^2 = 32.295, p < .001$).

Figure 5-1

COMPARISON OF "POINTER AND SETTER" DOGS
TO OTHER BREEDS OF DOGS ON EYE COLOR

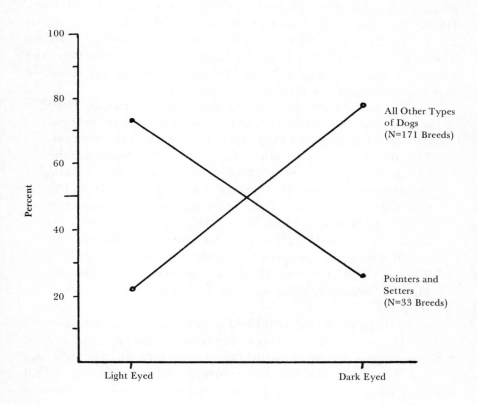

In addition to "Pointers and Setters" the only categories that included more light-eyed than dark-eyed breeds were "Retrievers" and "Spaniels". Retrievers were bred for the purpose of going into water or thicket to retrieve game that had been shot. Spaniels were bred as multiple purpose dogs. They worked partly like hounds and partly like pointing dogs. They were also trained to retrieve. All the other categories have more dark-eyed than light-eyed breeds and none of them are specialized in freezing-type behaviors as are the pointers and setters.

In addition to the above study of dogs which was done specifically with eye color in mind there have been other studies of dogs that, although done with other things in mind, can be reinterpreted in terms of eye color differences. Fuller and Clark (1966) compared the responses of wire-haired terriers (dark-eyed) and beagles (relatively light-eyed) in a controlled stimulus situation. The terriers were found to be much more spontaneous and responsive than the beagles. James (1951), who studied the same two breeds, summarized the differences in this way:

> The terriers are highly active and excitable animals. They would invariably come forward as the experimenter entered the runs and begin to jump up as if trying to get attention. The beagles, on the other hand, would remain in the background and would seldom come within reach of the experimenter's hand. If the terriers were removed from the group, however, some of the most aggressive beagles would come within reach. Most of the beagles were difficult to catch. They would run to the far end of the kennel and would have to be cornered before the leash could be applied. When cornered they would take the passive defense attitude. They are definitely of a more inhibited type than the terriers. (p. 71).

In 1958 Helen Mahut published a study done in Canada entitled "Breed Differences in the Dog's Emotional Behavior." She studied pets in their own home with the owner present. She was not interested in eye color, but her results can be reinterpreted in that context. To get a response she used a number of novel or frightening stimuli including a slithering mechanical snake, an umbrella, and a halloween mask. Responses were carefully noted and dogs that showed avoidance and wariness were categorized as "fearful." Those that were more bold in their reactions were categorized as "fearless." Of the ten breeds studied, seven breeds are light-eyed or variable in eye color, the other three breeds are always characterized as dark-eyed. The three dark-eyed breeds appear in Mahut's data as the top three breeds of the ten in fearlessness. Of 138 dogs in breeds with light or variable eye color, 84% were categorized as fearful; of 64 dogs in breeds that are always dark-eyed only 17% were categorized as fearful. This comparison is not as neat as we would like, since with eye color variable in some breeds, we do not know the eye color of the individual dogs involved. However, the magnitude of the difference between breeds is strongly suggestive of an eye color/reactivity relationship.

Behavioral Reactivity in Rodents

Two additional studies, done with rodents, are also relevant. The first was done by Tryon in 1931. Before starting a series of studies into the inheritance of maze-learning ability in rats, Tryon tried to determine whether any of his groups of rats already differed on this ability. One of the comparisons was between pigmented and albino male rats. Tryon stated, "The pigmentation variable consists of the two categories; presence and absence of pigmented eyes" (p. 18). He found that rats with unpigmented eyes were slightly but significantly better on the maze task. Tryon observed that the rats with pigmented eyes did less well because they tended to "be disturbed more easily by extraneous stimuli. By extraneous stimuli, I mean such as arise from movements of curtains and doors initiated by the animals themselves" (p. 19).

In other words, those with darker eyes did less well on this task because of a tendency to be overly reactive. To my knowledge, no one recognized the possible importance of that finding, which was published more than forty years ago, or even attempted to follow up on that early report of a link between eye-darkness and behavior.

Another study (Winston, Lindzey, and Connor, 1967) was done with pigmented and albino mice. This study beautifully demonstrates, in a laboratory setting, how animals with different inherited (reactive vs. nonreactive) tendencies, can use different behaviors to solve the same problem.

The problem presented the mice was one of avoidance learning. The apparatus used was a wooden box with an electrifiable grid floor. A platform (insulated and not electrifiable) was mounted above the grid floor. At floor level there was a round hole in each of the four walls through which the mouse could leave the box. Each trial of the experiment began by placing the mouse on the platform. Thirty seconds later the grid was electrified such that any mouse on the grid floor received an unpleasant shock. Initially, mice from both groups tended to jump from the platform, explore the grid floor, get shocked, and then escape through one of the holes. The purpose of the study was to see if, with repeated trials, the mice would learn to avoid the shock and if so how they would do it. There were two ways that the mice could avoid the shock. One way was to jump from the platform to the floor and go through one of the holes before thirty seconds had elapsed. The other way was to stay on the platform and not jump to the grid floor at all. The first strategy is one of active avoidance; the latter is one of passive avoidance. Either was an effective way to avoid the shock.

The results of the study were very clear. The two groups did not differ in their overall success at avoiding shock. However, they differed greatly in how they did it. The albinos almost invariably stayed on the platform and passively avoided the shock. The pigmented mice, on the other hand, were much more prone to jump off the platform, run through a hole, and thus actively avoid the shock. In discussing their finding, the authors state, "It seems probable that albinos have a greater inclination to freeze in the presence of aversive stimuli, while pigmented mice are more inclined to engage in some kind of motoric responding. There is previous evidence indicating that inbred Ss of the A (albino) strain tend to freeze or to be relatively inactive in a variety of situations involving aversive stimulation . . . " (p. 298). So again, although eye color was not mentioned as such, the two groups differed greatly in eye darkness and the findings were consonant with those already reported for other animals.

Generality of the Reactivity Model

The cumulative weight of the findings from so many different and varied forms of animal life and involving behaviors across a wide range of tasks indicates that the eye color/reactivity relationship involves a basic condition of nature rather than an isolated or chance covariation. The findings based on human performance are certainly more complex than are the findings based on animals, and caution is always in order when interpreting the causes of human behavior, but, taken as a whole the human and animal studies are consistent with each other. Indeed, from a Darwinian perspective it would be very strange if that were not the case.

Of course, the data on humans and the data on each of the animal groups involved specific behaviors that differ from one group to another. The behaviors are different, but the response requirements are similar in that they require either reactive or nonreactive skills. In each group we must assume that the reactive or nonreactive adaptation evolved because it was a superior adaptation in the given ecological situation.

For a group that evolved with the reactive adaptation, speed would appear to be more important than it would be for those with the nonreactive adaptation. Or, turning the statement around, one might say that for animals with good speed, the reactive adaptation would be good, but for slow animals the nonreactive adaptation would be better. The next chapter deals with speed of locomotion as a correlate of eye color.

References

Asdell, S. A. *Dog breeding.* Boston: Little, Brown and Company, 1966.

Davis, H. P. *The modern dog encyclopedia.* Harrisburg, Pennsylvania: The Stackpole Company, 1956.

Fuller, J. L. and Clark, L. D. Genetic and treatment factors modifying the postisolation syndrome in dogs. *Journal of Comparative and Physiological Psychology,* 1966, *61,* 251-257.

James, W. T. Social organization among dogs of different temperaments, terriers and beagles, reared together. *Journal of Comparative and Physiological Psychology.* 1951, *44,* 71-77.

Mahut, H. Breed differences in the dog's emotional behavior. *Canadian Journal of Psychology,* 1958, *12,* 35-44.

Sneider-Leyer, E. *Dogs of the world.* New York: Arco Publishing Company, 1970.

Tryon, R. C. Individual differences in maze ability. II The determination of individual differences by age, weight, sex and pigmentation. *Journal of Comparative Psychology,* 1931, *12,* 1-22.

Winston, H. D., Lindzey, G., and Conner, J. Albinism and avoidance learning in mice. *Contemporary Research in Behavioral Genetics,* 1967, *63,* 77-81.

Chapter 6

Eye Color and Speed of Movement

As it became clearer that dark eyes were associated with reactive skills, the question arose whether dark eyes were also associated with superior speed of locomotion. The answer cannot be given conclusively, but there is some limited evidence which suggests such a relationship.

For anyone who follows track performance in the United States, it is clear that blacks excel as sprinters. The domination is overwhelming and has been clearly evident for many years. There are many interesting interpretations advanced to explain the superiority—some cultural and some biological—and probably more than one factor is involved. That blacks tend to run faster than whites is true not only of athletes, but, as noted by Kane (1971), it is also true of college non-athletes and grade school children.

In addition to the studies that compared blacks and whites, there was also an important study by Codwell (1949) of black high school boys who were classified as "Dominantly Negroid", "Intermediate", or "Strong Evidence of White". One of the criteria used for classification was eye color. Of the ability measures tested by Codwell one was sprinting speed. Those students classified as "Dominantly Negroid" were significantly faster sprinters than were those students in the "Strong Evidence of White" group.

Another area which presents suggestive evidence is professional football. As pointed out in Chapter 2, blacks and dark-eyed whites tend to play those positions that depend most on speed. Blacks and dark-eyed whites are more likely to play defensive back than any other position and that position, perhaps more than any other, demands speed.

Other than the few human studies there is very little information available about eye color and speed among mammals. Among wild mammals there are both light-eyed and dark-eyed species that are noted for speed. If there is a relationship—positive or negative—

63

between eye color and speed for wild mammals it is not apparent from readily available information.

With dogs, the relationship between eye color and speed is uncertain. There is one group of dogs, the coursers, that were bred for speed. They were developed to use together with pointers in hunting rabbits and other game. After the pointers located the game, the coursers were brought forward to give chase. The various breeds of coursers do tend to be dark-eyed which contrasts them with sporting dogs such as pointers, setters and retrievers, but since most breeds of dogs are dark-eyed the coursers do not apparently differ from dogs in general.

When we extend our consideration to lower forms of animal life, the suggestion of an eye color/speed relationship becomes stronger. Among the birds of prey, for instance, those most noted for speed, the falcons, are also much darker-eyed than other birds of prey. Additional evidence comes from a study I did based on published records of flight speed for different species of birds. The source used was an article published by Tucker and Schmidt-Koenig (1971) in *The Auk,* a journal of ornithology. They had carefully measured horizontal flight speed for 21 species of birds—mostly shore birds. By using other sources it was possible for me to classify the various species as light-eyed or dark-eyed. The speeds ranged from 11 M.P.H. to 21 M.P.H. The average speed for the nine light-eyed species was 12.8 M.P.H.; the average speed for the twelve dark-eyed species was 15.4 M.P.H. This difference in average speed favoring the dark-eyed birds was statistically significant ($t = 2.20$; $df = 19$; $p < .05$).

Moving to a lower phylogenetic level, we find an interesting example with snakes. Most snakes are not known for speed of locomotion, but there is one group of snakes so fast that they are called racers. Interestingly enough, the racers differ from other snakes in having a large black blotch of pigment on the iris which makes them darker eyed than most snakes. That greater speed and dark eye blotches occur in the same type snake may be coincidental, but perhaps not.

Moving on to a lower level, there are two good studies that suggest an eye darkness/speed relationship in insects. The first was a study published by Scott (1943) in the *American Naturalist.* Scott worked with fruit flies of different eye colors. He used three strains that clearly varied in eye darkness—a brown-eyed group, a red-eyed group, and a white-eyed group. Fruit flies have the inherited tendency to move toward a light (i.e., they are positively photo-tropic). In Scott's study a light was suddenly presented and speed of crawling toward the light was measured. The actual measure was the time in seconds it took the flies to crawl a distance of about 18 centimeters—an 18 centimeter dash, if you will. The results indicated a linear relationship between eye darkness and speed; the brown-eyed flies were fastest, the red-eyed flies were next, and the white-eyed flies were slowest.

The other relevant insect study was done with locusts. Because locusts and grasshoppers do so much agricultural damage they have been extensively studied by entomologists. Uvarov (1966) has summarized this research. Of interest here is the fact that some types of locust go through stages of light and dark pigmentation. The changes in pigmentation are not limited to the eyes, but do include the eyes. During the period of light pigmentation the locusts are relatively inactive and are solitary. During the pigmentation of dark pigmentation they are active and gregarious. Comparative studies have been done to determine other differences in behavior from one stage to another. Since locusts tend to engage in "marching" behavior, controlled studies have been done to measure marching speed during each of the stages and statistically significant differences in marching speed were found. Locusts in the darkly pigmented stage marched faster than locusts in the lightly pigmented stage.

Summary

To summarize, the evidence is spotty, but there is some reason to believe that dark-eyed organisms are not only more reactive but also faster in locomotion than light-eyed organisms. The available information indicates the probability of such a relationship, but, except for some suggestive findings with humans, most of the evidence involves forms of animal other than mammals. We really cannot say for sure at this point whether or not there is a general tendency in nature for eye pigmentation and speed of locomotion to covary. If there is such a general relationship, however, there is little doubt about the nature of the correlation; darker eyes are positively associated with greater speed.

So, to review Part Two, we found strong evidence of an eye darkness/reactivity relationship from such scattered areas as human athletic performance, hunting and escape tactics of birds and mammals, behavioral traits in different breeds of dogs, and laboratory studies of rodents. We also have some reason to believe that speed of locomotion goes with the reactive adaptation. Hopefully,

65

much research will be done in the future to further test the validity and generality of the eye darkness/reactivity relationship. At this point, however, the phenomenon has been sufficiently demonstrated to warrant theoretical speculation and inquiry into possibly related areas of behavioral and physiological functioning. In the chapters to follow I will attempt to organize information from diverse sources along these lines. Some original research will be presented, but most of the empirical findings are those of others.

In Part Three some relevant perceptual and physiological processes will be discussed. The functions of the eye will be reviewed and the functional importance of eye pigmentation will be considered. Recent studies linking eye darkness to differences in visual perception will be reviewed. Non-visual functions of the eye will be discussed in terms of findings by physiologists and endocrinologists about the influence of light on biological functioning.

In Part Four additional differences between light-eyed and dark-eyed organisms will be discussed. These differences will be shown to appear in diverse areas of human and animal function and dysfunction.

References

Codwell, J. E. Motor function and the hybridity of the American Negro. *The Journal of Negro Education,* 1949, *18,* 452-464.

Kane, M. An assessment of 'black is best.' *Sports Illustrated,* 1971, *34,* (3), 72-83.

Scott, J. P. Effects of single genes on the behavior of Drosophilia. *American Naturalist,* 1943, *77,* 184-190.

Tucker, V. A. and Schmidt-Koenig, K. Flight speeds of birds in relation to energetics and wind directions. *The Auk,* 1971, *88,* 97-107.

Uravov, B. *Grasshoppers and locusts,* Vol. 1, Cambridge: At the University Press, 1966.

Part Three

The Eye and Biological Functioning

Chapter 7

The Role of Eye Pigmentation
in Vision and Light Sensitivity

Earlier chapters have demonstrated the consistency with which certain behaviors are associated with eye color. Eye color, being a physical trait familiar to everyone, has not been dealt with other than to describe the bases used for rating eye-darkness. As we seek to gain more understanding of why or how eye color is related to behavior it is necessary to look more closely at eye color and other aspects of the eye as well. That information will give the problem a physiological context which will be useful as we proceed. For this information I have relied heavily on several very good sources (Fox and Vevers, 1960; Walls, 1963; Marler and Hamilton, 1966; and Droscher, 1969).

Light and the Visual Spectrum

Light has been a central element in the environment of evolving organisms. When we speak of light we are usually talking about that part of the total spectrum of radiant energy that can be perceived by animals. That part is only a small proportion of the total spectrum and interestingly enough is composed of those wavelengths that most readily penetrate water. This fact is most likely related to the aquatic origin of life. Cosmic rays, gamma rays, X rays, and ultraviolet rays are shorter than those perceived. Rays that are of longer wavelengths than those perceived by living organisms include infrared, heat waves, spark discharge, radar, radio, and slow electromagnetic waves. The visual spectrum is bounded on the shortwave end by ultraviolet waves and on the long-wave end by infrared.

The visual spectrum (perceived light) is that part of radiant energy with wavelengths that range from about 400 millimicrons to 750 millimicrons. Reflected light of those wavelengths are the ones perceived, but they are not necessarily perceived as alike. Some animals, including man, are capable of differentiating among reflected lights of different wavelengths. What is perceived, of course, is not different wavelengths, but different colors. The colors and their approximate limiting wavelengths are as follows:

Color	Wave Length Millimicrons
Red	605 — 750
Orange	590 — 605
Yellow	575 — 590
Green	490 — 575
Blue	450 — 490
Violet	400 — 450

The numerous other colors we perceive are shades between or mixtures of the above colors. Purple, which is a mixture of the two extremes, violet and red, allows one to think of the visual spectrum as circular with each color shading into the one next to it. The color of an object depends on the wavelengths reflected by that object. Which wavelengths will be reflected depends not only on the chemical composition (nature of the pigments present) of the object, but also to some degree on the physical structure of the object. If an object reflects all wavelengths it appears white; if it absorbs all wavelengths it appears black. A mixture of black and white particles appear gray.

Looking again at the chart of colors matched to wavelengths one can see that the various colors do not represent equal proportions of the visual spectrum. For instance, the range of wavelengths seen as red is broad; the range of wavelengths seen as yellow, on the other hand, is relatively narrow. Colors such as red and green that cover a broad range of wavelengths occur more often in nature than do those such as yellow and orange, that cover narrow ranges of wavelengths.

Although we have referred to this as the visual spectrum, in fact there are many visual spectrums slightly different in their ranges. Some animals are blind to wavelengths seen by us, but may see wavelengths that we do not. For example, a bee can not see red, but it can see ultraviolet colors not seen by man. There may also be slight differences in the visual spectrum from one member to another of the same species. So the visual spectrum may vary from one species to another or from one organism to another, but its overall range is relatively constant. The longwave end of the spectrum is referred to as the red end and, although not exactly accurate, it is common practice to refer to the shortwave end of the spectrum as the blue end.

The Eye as a Living Camera

Light reception and utilization is almost ubiquitous in animals. Even the tiniest organisms are equipped to receive light. Through time animals have evolved organs specialized for light reception. Such an organ is the human eye. Its best known function is that of vision and its basic parts have often been compared to those of a camera. Like most analogies it is far from perfect, especially when functional specifics are considered, but it does help to get in mind some of the structural elements. The match of similar elements is as follows:

Eye	Camera
Retina:	Film or plate
Cornea and lens:	Lens
Lids:	Shutter
Pupil:	Aperature—opening for light to enter
Iris:	Diaphragm that controls the size of the opening

If we think then of the eye as a camera with the shutter usually open except when the organism is sleeping we recognize that the center of the eye as we look at it is an opening, the pupil. The opening is covered by a transparent structure, the cornea. Light which passes through the pupil is focused by the lens on a layer of cells, the retina. The size of the opening (pupil) determines how much light is allowed into the eye. Changes in the size of the opening are controlled by the iris which can contract or dilate to make the pupil smaller or larger. As we look at the eye, we see the pupil is in the center with the iris around it. The iris is the part of the eye which varies in color. When we say eye color we are talking about the color of the irises.

The retina, which has the job of processing and transmitting the input of light signals, is a complex structure consisting of several layers of cells. The cells are typical of brain cells and the retina is considered a part of the central nervous system. It may be a mistake, however, to think of the eye as an outgrowth of the brain as is often popularly suggested. It would be more accurate to think of the brain, or at least part of it, as an evolutionary extension of the eye.

Melanin and Its Functions

Continuing our analogy of an eye as being somewhat like a camera, we note that a camera has the form of a black box with a hole in it. The box is made light-proof and painted black to ensure that the light which enters the camera through the tiny opening will contrast sharply with the surrounding area of the film and allow a clear image. The eye likewise is relatively light-tight except at the pupil. The material which serves as the light shield for the eye is the pigment, melanin.

Melanin is not the only pigment found in animals, but it is the most important one and the one with which we are primarily concerned. Melanin may vary somewhat in color under certain conditions but usually it is brown; however, if the granules of pigment are dense the color appears black. Indeed the word "melanin" comes from a word that means black. Melanin occurs in branching cells similar in form to nerve cells; such cells are found in the hair and skin as well as in the eye. A chief function of melanin is to serve as a light shield, especially against ultraviolet (short wavelength) light. For instance, it is known that light-skinned persons, having a relatively small amount of melanin in the skin are very prone to skin cancers caused by excessive exposure to sunlight.

Melanin and Eye Color

Melanin occurs in the eye in such a way as to surround the retina and act as a light shield. There is a layer of pigment behind the retina which we do not see except to the degree that it makes the pupil, which is only an opening, appear black. The melanin pigment that we do see clearly is in the iris. As noted earlier, it is the iris which varies in color. There are several layers of the iris which may be pigmented with melanin. In normal Africans all layers are heavily pigmented; in albinos on the other hand little or no pigmentation occurs in any layer of the iris or in the retina either.

Differences in the amount of melanin found in the iris vary between those extremes. Differences in eye color are directly related to the amount and distribution of melanin in the iris. If melanin is brown, then one might question at this point how eye color could be anything other than brown or black. The answer is that the color we perceive is not a simple reflection of the color of the pigments involved; perceived color is also dependent on certain optical effects. If tiny dark particles are viewed against a black background, light of

72

different wavelengths is scattered in such a way that the color perceived is blue. This scattering effect, which is called Tyndall scattering, makes the sky appear blue and some eyes appear blue. Blue eyes, then, are those that have only tiny particles of pigment in the outer layers of the iris. With particles of increasingly larger size the eye color changes from blue to green, to hazel, to brown, to black. So, although we tend to think of eye colors as discrete, in fact, they are as continuous as height or weight, depending on the amount of melanin present.

The Inheritance of Eye Color

Physical traits that are continuous rather than discrete depend on multiple pairs of genes that add together to give an overall effect. The continuous and polygenic nature of eye color has often not been recognized, a fact pointed out by Moody (1967):

Some readers may be surprised to find this subject included in a discussion of polygenic inheritance. If there is one thing the average person thinks he knows about human inheritance, it is that "brown eyes are dominant to blue." Traditionally this has been regarded as the best human example of Mendelian inheritance based on a single pair of genes, brown-eyed people having the genotypes BB or Bb, blue-eyed people the genotype bb. Such simplicity is appealing, yet a little thought will convince us that we should not really expect it to be true.

In the first place, people cannot really be classified so easily as brown-eyed, or blue-eyed. As we know from observation, eye colors come in all sorts of shades that are more or less intermediate between brown and blue—hazel, green, gray, light brown, dark brown, black. This range of variation in itself suggests polygenic inheritance (p. 171).

Moody goes on to review genetic studies that have demonstrated empirically that eye color is a continuous variable depending on multiple pairs of genes. The number of genes involved has not been determined with certainty, but the data suggest that some of the genes are sex-linked and carried on the X chromosome. Since females have two X chromosomes and males only one, recessive traits carried on the X chromosome are much more likely to occur in males than in females. If the recessive gene involved is only one of many that determines the trait (i.e., the trait is polygenic), males will differ only slightly from females in overall degree of the trait. This seems to be the case with eye color; females tend to have slightly darker eyes than do males. This has been noted in most human societies that have enough light-eyed people to allow easy comparison. It seems probable that decreased iris pigmentation is carried as a recessive trait on the X chromosome. Again though, it must be stressed that there are several genes involved in determining eye color.

Albinism

If an organism is severely lacking in pigmentation it is referred to as an albino. Albinism, which is an inherited characteristic is found in fish, amphibians, reptiles, birds, and mammals, including humans. It is found in all human races but is more common in some populations than in others. The frequency, for instance, is very high among Hopi Indians. Total albinism appears to involve a deficiency in the enzyme, tyrosinase, which is necessary for the synthesis of melanin.

There are several types of albinism, each with its own characteristic mode of inheritance. Albinism may be total, involving depigmentation of the skin, hair, and eyes, but it is also possible to inherit each separately. Cutaneous albinism is inherited as a dominant trait. Ocular albinism is inherited as a recessive, sex-linked trait (Fitzpatrick and Quevedo, 1966). It is not uncommon for relatives of persons with ocular albinism to also have irises that are below normal in pigmentation. This suggests that they are heterozygotic for the trait and experience it in less than full force. For persons with total albinism or total ocular albinism there is little or no pigment in the iris. The eyes of many albinos do not appear blue, however, but pink. The reason for the pink appearance is that since there is melanin neither in the iris nor behind the retina, the pigments nearest to the surface are those of red blood. The amount of pigmentation present in an albino is related to the normal level of pigmentation in his racial population. Thus an African Albino may have eyes that range in color from blue to light brown.

Human albinos were once called "moon-eyed" because they saw better at night than in the day. Marked photophobia is noted in both human and animal albinos. This increased sensitivity to light is accompanied by impaired vision. Studies have shown that the amount of impairment is related to the amount of depigmentation.

Two Functions of the Eye

Whenever we think of the eye, we tend to think of vision. Vision, however, is not the only function of the eye and was certainly not the first function from an evolutionary standpoint. Light has been an everpresent and powerful element in the environment in which life has evolved. Light is also an element that continuously changes in intensity. These changes are cyclic and follow consistent daily, monthly, and yearly patterns. Life evolved the capacity to respond to those changes prior to evolving the capacity for vision. Even rudimentary forms of plant and animal life are capable of responding to changes in light.

As animals evolved, the light reception function became specialized. That is, some cells were more sensitive to light than others and through the evolutionary process these cells became more and more specialized and more and more efficient as **photoreceptors**. Efficiency in this case involved sensitivity to weak intensities of light and, perhaps to some degree, differential sensitivity to different wave lengths.

Sensitivity to light is important to plants and animals because rhythms of metabolism and other bodily functions are synchronized and directly influenced by light. For many such functions, that **photoreceptor** (eye) is most efficient which is most sensitive.

Very well, sensitivity is an asset, but only for the first function of the eye. For the other function of the eye, vision, resolving power is the needed characteristic. In the human eye, that which increases sensitivity decreases resolving power and vice versa. Walls (1963) is very emphatic on this point, "Sensitivity and resolving power are thus on the two ends of a seesaw, and whatever sends one up sends the other down" (p. 89). At another point he stated, "They are indeed so very different that they are practically mutually exclusive" (p. 65).

These two functions are related to eye color because iris pigmentation retards sensitivity and facilitates resolving power. Since melanin acts as a light shield, reduced pigmentation allows more light into the eye in the area surrounding the pupil (or on the retina in the area surrounding the fovea). This additional reception of light makes the eye more sensitive but at the cost of reducing the contrast needed for visual acuity. Each species, as it evolved, was forced to make adaptations that represented some compromise between these two contrasting needs.

75

Adaptations for Light Reception

Each species has evolved toward better adaptation to its light environment according to the pressures of natural selection. The particular structures that evolved to meet similar needs were often quite dissimilar.

One of the oldest adaptations, common in insects, is multiple eye-spots or ocelli. These spots, which may be anywhere on the body, are sensitive to the absolute level of light. In some primitive organisms eyespots are the only organs for light reception; in others eyespots supplement the compound eyes used in vision. Bees, in addition to their compound eyes, have three tiny ocelli. They are between the compound eyes and are useful to the bee in making fine discriminations at dawn and dusk as to when to make the first and last trips of the day. In some primitive animals, several eyespots working together provide a kind of crude vision. Compound eyes as we know them undoubtedly evolved from similar primitive eyespots.

Extra eyes are not always, however, just crude "light meters" reading the absolute level of light. Many fish, amphibians, and reptiles have a more complex "median eye" or "third eye" (some have two) that, in some cases, is capable of making wavelength discriminations. This eye occurs in a median or central position above the compound eyes and may have a very simple lens and retina. It has a nerve running to the pineal gland in the brain and is often called the "pineal eye". The horseshoe crab has two such eyes that are used to register ultraviolet light. This capacity is valuable to the crab in making light discriminations in cloudy weather. An even more fascinating use of this eye is made by lizards:

> But most true lizards also have a median eye. Its function is amazing; depending on the colour of the light which reaches the pineal gland through this eye, the gland produces more or fewer hormones; and these in turn cause a change in the reptile's skin so that after only a few minutes its colour resembles that of the light, producing an ideal camouflage (Droscher, 1969, p. 50).

76

Another adaptation related to need for light is size of the eye. Nocturnal animals and deep sea fishes, having a need to detect faint light, have large eyes. The eye of the owl has evolved to a size that is too large to move in its socket. Since the eye can't be turned, the owl has to turn the entire head in order to look in a different direction. Likewise, animals with a great need for light sensitivity tend to have large pupils relative to animals with a great need for visual acuity. Some adjustment of pupil size is possible, of course, for most animals.

Another of nature's ways of increasing eye sensitivity in certain animals, notably cats, has caused us at times to think an animal's eyes were shining in the dark. A cat's eyes will often appear to glow at night when seen from an automobile because the eyes are reflecting the light from the headlamps. This occurs because cats and some other nocturnal animals have a mirror-like lining, the tapetum lucidum, at the rear of the retina which reflects light back through the eye giving the retina a second chance to register the light input. This increases sensitivity in dim light at the cost of reducing visual acuity in bright light.

For animals that need lowered sensitivity because of exposure of the eye to bright light, an adaptation which has evolved is a colored lens. Ground squirrels and prairie dogs have lenses that are yellow or orange in color. Among snakes, the color of the lens is darkest in those species, such as the racers and whip snakes, most exposed to bright light. These snakes, as noted earlier, are also noted for speed of movement.

The darker or more red-like the coloration of the lens, the more the lens blocks light of short wavelengths. In that respect, a colored lens is similar to iris pigmentation in blocking more short wavelength than long wavelength light. This is illustrated well by changes in color that occur in the lens of the human eye. With age the human lens gets steadily more yellow. As a result, whereas the lens of a child blocks only about 10% of blue light entering the eye, the figure may be higher than 80% for people in their 70's or older. As a result, elderly artists often have trouble making fine distinctions when working with blue paint. If the lens is surgically removed the person becomes very sensitive to violet and even ultraviolet colors not normally seen by human beings (Gates, 1946, p. 134).

Another adaptation, this one common in birds, involves the secretion by the organism of colored eye droplets. In this way, the

77

bird gets the benefit of a colored filter which has the same effect as a colored lens, but the droplet, unlike the colored lens, is a transitory condition. As the light conditions change, the amount and color of eye droplets also change. In bright light the secretion of red droplets facilitates vision; in dim light the absence of droplets or droplets of a different (shorter wave-length) color facilitate sensitivity to changes in illumination. The advantages of this type of adaptation is that the animal is able to enjoy the benefits of bright light acuity *and* dim light sensitivity. The crayfish accomplishes the same thing by means of a system whereby migration of melanin changes the amount of eye pigmentation present at night from the amount present during the day.

Flexibility is also achieved in yet another adaptation—that of different type cells in the retina for different functions. In most animals, including man, the retina contains a mixture of rods and cones. The cones, which respond more to long wavelength and bright light, are concentrated in a central portion of the retina called the fovea. The rods, which are more sensitive to short wavelength and dim light are located primarily in the periphery of the retina. The human eye contains about 125 million cones and about 6 million rods. When exposed to bright light the eye becomes light-adapted and the cones are maximally active; when exposed to darkness the eye becomes dark-adapted and the rods are maximally active. An interesting aspect of dark adaptation is that, since rods are not concentrated in the center of the retina as are the cones, a person can see best in the dark by looking a little to the side of the thing he wants to see.

As the above discussion of various adaptations makes clear, amount of ocular pigmentation is just one variable that affects the sensitivity and resolving power of the eye. It is a most important one, however, for the study of human behavior because it is the only one in which there are large known differences between human individuals and groups.

Habitat and Eye Color

In reading about the habits of American birds I had gotten the impression that birds with light eyes would be more likely than birds with dark eyes to live on the seacoast, where it is often cloudy, or in dense vegetation. In order to test the latter observation, I tallied information on the land birds of India (Kinnear, 1949) and noted that some species were identified as living in "dense" habitats. I recorded any reference to dense habitat and eye color for 242 species of land birds. A significantly greater percentage of light-eyed (Ratings 1 to 3) than dark-eyed (Ratings 4 and 5) species were so identified. ($X^2 = 6.222$, $df = 1$, $p < .02$)

The observation that light eyes go with dense habitat is true for broad categories of "dark eyes" and "light eyes" but does not differ among the three levels within the "light-eyed" group. This can be seen in figure 7-1.

Our information on human eye color and habitat is limited but worth noting. Most human beings around the world have dark eyes. Among northern Europeans and their descendants, however, light eyes are in the majority. The highest percentage of light eyes is found around the Baltic Sea. That area is characterized by frequency of heavy cloud cover. This aspect of the environment may have been an important factor in permitting the light-eye mutation to survive.

Light eyes in humans have evolved together with lightly pigmented hair and skin. The three traits are correlated positively with each other but the correlations are far from perfect. Many people with dark hair or skin have light eyes and many people with light hair or skin have dark eyes. Some Eskimos have light eyes as do some American Blacks. Light eyes have been reported by Beals and Hoijer (1965) to occur ". . . among the offspring of blue-eyed Europeans (or Americans) and brown-eyed Japanese, where the offspring, though displaying a number of characteristics like the Japanese parent, also show a surprising frequency of blue eyes" (p. 165). As people continue to migrate and intermingle any relationship between habitat and eye color in humans will probably become progressively weaker.

Temporal Changes in Eye Pigmentation

We tend to think of eye color as a constant physical trait that does not change from time to time. For most practical purposes that is true, but there are some conditions under which eye color undergoes slight changes. First of all, eye color changes with age. Most babies are born with blue eyes that gradually darken with age. Again in old age, eyes become lighter. The change in pigmentation of hair and eyes with age is apparently influenced by changes in the level of sex hormones. That pigmentation is influenced by various

Figure 7-1

EYE-DARKNESS AND TENDENCY TO LIVE IN
A DENSE HABITAT FOR SPECIES OF LAND BIRDS
IN INDIA

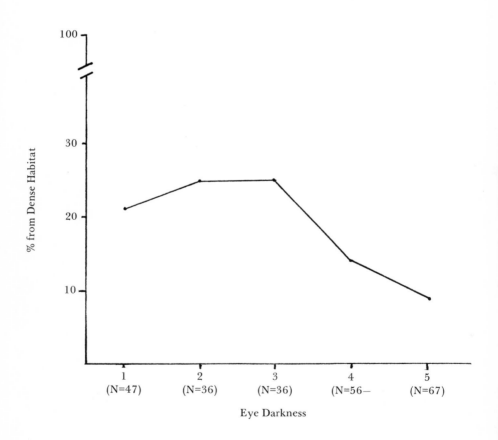

hormones has been demonstrated in numerous species (Snell, 1967). Slight changes in skin pigmentation of human females have been observed to vary with changes in the menstrual cycle. Winge (1950) reports that the eyes and nose of female dogs darken toward the end of pregnancy. Eye color of certain birds also darkens slightly during the breeding season.

Eye pigmentation is also influenced by hormones other than sex hormones. Stress conditions that increase adrenal output have the effect of slightly, but noticeably, darkening the eyes of some animals. When an Eagle Owl is made angry the iris color changes from yellow to red; when a Rock Bass is given electric shock the iris likewise reddens (Walls, 1963). The iris may also be darkened by simply giving injections of adrenaline. So fluctuations in eye color do occur, but they are relatively minor. As a general rule, eye color may be thought of as a highly stable physical characteristic.

References

Beals, R. L. and Hoijer, H. *An introduction to anthropology*. New York: The Macmillian Company, 1965.

Droscher, V. B. *The magic of the senses*. New York: E. P. Dutton, 1969.

Fitzpatrick, T. B. and Quevedo, W. C., Jr. Albinism in J. D. Stanburg, J. B. Wyngaarden and D. S. Fredrickson (Eds.). *The metabolis basis of inherited disease*. New York: McGraw-Hill Book Company, 1966, 324-340.

Fox, H. M. and Vevers, G. *The nature of animal colors*. London: Sidgwick and Jackson Limited, 1960.

Gates, R. R. *Human genetics*. New York: Macmillan, 1946.

Kinnear, N. B. *Whistler's popular handbook of Indian birds* (4th ed.) Edinburgh: Oliver and Boyd, 1949.

Marler, P. and Hamilton, W. J. III *Mechanisms of animal behavior*. New York: John Wiley and Sons, 1966.

Moody, P. A. *Genetics of man*. New York: W. W. Norton and Company, 1967.

Snell, R. S. Hormonal control of pigmentation in man and other mammals, in W. Montagna and F. Hu (eds.) *Advances in biology of the skin*, Vol. 8, *The pigmentary system*. Oxford: Pergamon Press, 1967, 447-466.

Walls, G. L. *The vertebrate eye and its adaptive radiation*. New York: Hafner Publishing Company, 1963.

Winge, O. *Inheritance in dogs*. Ithica, New York: Comstock Publishing Company, 1950.

Chapter 8

Eye Pigmentation and Perception

In the preceding chapter the eye was considered as an organ with competing functions. The suggestion was made that eye pigmentation helps determine the point of balance between the competing functions. An eye with heavy pigmentation would have heightened visual acuity but lessened visual sensitivity. Likewise, since melanin blocks short wavelength (violet or blue) light more than long wavelength (red) light, it was suggested that organisms with dark eyes should be relatively more responsive to colors at the red end of the spectrum and light-eyed organisms more responsive to colors at the blue end.

In this chapter we will explore further the evidence to support these suggestions and try to point out how differences in perceptual ability are related to the life of the organism. Our focus will change from the eye as such to those behaviors related to perception. First we will deal with observations and studies based on animal behavior and then review supporting evidence from recent studies of human perception.

Observations from Nature

An animal must make appropriate responses to survive. Some responses are acquired through learning, and some through genetic selection. What constitutes an effective survival response depends on a complex and ever-changing ecological system. The evolutionary direction taken by a particular species depends to some degree on the direction taken by other species of plants and animals. Some plants, such as grasses, for example, are pollinated sufficiently by the movements of wind and water. These plants have not developed attracters such as brightly colored flowers. Plants that are pollinated by insects or birds have ways of attracting these animals. Those plants have survived whose flowers attract the available animals; in turn, among animals that depend on flowers for food those have survived that were attracted to the available flowers.

Plants evolved flowers with the color most attractive to available animals and animals evolved visual preference for the flower colors that most adequately met their survival needs. This process has resulted in flowers of varied colors and pollinating animals of varied color preferences. Bees and moths are drawn to blue or violet flowers since they are more sensitive to short wavelength colors than are most animals. They cannot see red as a color at all and are not attracted to red flowers. Hummingbirds and butterflies, on the other hand, can see long wavelength colors and are more attracted to red and yellow than they are to colors of short wavelength (Grant, 1959a; Grant 1959b). To ask which came first, the animal's preference for red or the availability of red flowers, is somewhat like asking which came first, the chicken or the egg—the answer is they evolved together.

Not all color preferences depend on mutually supporting and mutually evolving species. Frogs only see blue as a color and have an instinctive attraction to blue. When in danger they jump toward blue (water) rather than green (grass) (Droscher, 1969). Frogs evolved a preference for blue because that is the color of water and more jumpers-to-water than jumpers-to-grass survived to have offspring (we do not assume that water evolved a blue color in order to attract frogs).

Jumping into water (blue) when excited evidently has survival value for the frog, but not all color preferences of animals can be traced to clearly evident causes. Minks, for instance, are drawn to red and this fact is used to trap them (Walls, 1963); many hawks appear to like green and often bring fresh leaves to the nest (Brown and Amadon, 1968). These preferences may have some subtle survival value, so far undiscovered, or they may be merely a by-product of eye characteristics, such as amount of eye pigmentation, that evolved to their present state for reasons having nothing to do with color preference. For many species of animals color perception and color preference apparently play little part in their present existence.

Most mammals are color blind. Only the primates and a few other species (including squirrels and perhaps cats) can discriminate colors. On the other hand, birds, lizards, turtles, frogs, and teleost fishes have all been demonstrated to have color vision. Why most diurnal mammals do not share the same capacity is a major puzzle of evolution (Gordon, 1968, p. 451).

When it comes to perception of color, we primates are more like birds than we are our fellow mammals. The bird visual spectrum is fairly similar to our own, and colors, in their life as in ours, play an important part. Consequently there is a source of readily available information on birds which can be used to test the hypothesis that dark-eyes are more responsive to long wavelength colors than are light-eyes. It has already been mentioned that hummingbirds are especially drawn to red and yellow flowers and hummingbirds are darker-eyed than most birds. That particular observation is consistent with the hypothesis, but taken alone it does not mean very much. We need to make comparisons between light-eyed and dark-eyed cases to conclude that something more than chance is operating. Fortunately such comparisons can be made.

Color, especially of the breast feathers, is very important to many birds in attracting a mate. Darwin pointed out long ago that natural selection would favor those birds with the colors most preferred in sexual selection. Many species have evolved bright breast colors, especially the long wavelength colors of red and yellow. It is reasonable to infer that birds that have evolved these colors did so because they are more sensitive to them than are other birds. Logically, it is just as proper to suppose the other direction of causality (i.e., feather color came first and color preference followed). At any rate, a consistency between the two would tend to have survival value.

If dark-eyed birds are more sensitive than light-eyed birds to long wavelength colors we might expect that more of them would have red or yellow breast colors. I used the Pearson (1936) work, *Birds of America,* to test that hypothesis. For each species, the colors listed under the heading "Under Parts" were tallied. This involved a total of 717 color references. Of the total, 545 were from descriptions of dark-dyed birds and 172 from descriptions of light-eyed birds. For the dark-eyed birds 23% of all references were to colors in the red-yellow range. The figure for light-eyed birds was only 13%.

The results might be questioned on the basis that eye color and feather color are biologically closely related. Thus, any change in one might have the effect of changing the other. In that case the two would be correlated, but one would not be the cause of the other. To be clear of that criticism, a situation is needed for study in which the bird's eye color and the other colors involved are clearly and completely separate in origin. Such a situation, fortunately, does exist in nature.

In the wilds of New Guinea and Australia live a family of birds that are truly fascinating. These are the bower birds. They are called bower birds because the males build "gardens" or "playgrounds". These playgrounds or bowers are often quite elaborate. First of all the bird clears the area of all debris. Then it builds a platform of grasses which covers about four square feet. It then constructs two walls to form a "hall" or walkway leading to the platform. The walls, which are about one foot long and one foot high, are made by inserting sticks into the ground. Some species even paint the inside of the walls. The stain they use comes from charcoal made wet by saliva. The bird chews on fibrous bark to make a "brush" with which to dab the walls. The walkways (like most football fields and perhaps for the same reason) always run north-south rather than east-west (Iredale, 1950). The exact pattern of the bower varies with different species and there is some uncertainty about the function of the bowers. Probably they serve to attract females, but some observers have questioned this, at least for some species, since the bower is not placed near the nest and females are seldom observed at the bower. It is very easy to anthropomorphize and think of these male birds as building and maintaining the bowers, at least partly, for their own entertainment and pride of ownership.

Ownership extends beyond just the structures themselves. On the platform the bird places all its valuables. They are sorted according to color if more than one color is involved. These valuables may consist of numerous kinds of objects—pieces of paper, leaves, berries, buttons, shells, coins—as long as they are the desired color(s). Different species prefer different colors and this fact can be related to eye color. There are not very many species of these birds but they do, luckily, cover the full range of eye colors. Of the twelve species for which we have color preference information (Gilliard, 1969) eye color ranges from light blue to dark brown. Since both eye color and color preference (as indicated by color of objects collected) could be quantified, I attempted to see if the two were correlated. Eye darkness ratings were assigned to different colors as before on the same five point scale. The hues mentioned for each species as ones preferred were also rated as to wavelength. A simple rating was used as follows: red = 4, yellow = 3, green = 2, and blue = 1. The ratings were in order of wavelength—the longest wavelength color (red) was given the highest rating and each of the progressively lower ratings indicated shorter wavelength. Preference for black, brown, white and gray were not included since they are not true colors in the same sense as the others. For each species, then, there were two values: (1) rated eye-darkness and (2) an averaged rating of all colors preferred by the species as indicated by color of objects collected. It was expected that darker-eyed birds would prefer more reddish objects and light-eyed birds would prefer more bluish objects. Such was the case; the correlation between eye darkness and color of objects collected was highly positive and statistically reliable (Product-moment correlation = .75, $n = 12$, $p < .01$).

If bower birds are fascinating and worthy of study, the species to be considered next is no less so.

Eye Color and Human Perception

There is ample evidence that dark-eyed people have better visual acuity than light-eyed people. Albinos, whether human or animal, have poor vision. There are indications that the deeper the coloration of the albino eye the better the visual acuity (Falls, 1953). A study of visual acuity in normal American soldiers (Karpinas, 1960), reviewed by Dreger and Miller (1968), involved more than a quarter of a million men. It was found that 82% of the black soldiers as compared to only 69% of the white soldiers had 20/20 uncorrected distant vision. A study done in Europe iuvolving more than two thousand children found that light-eyed children had more visual defects than did dark-eyed children (Bassin and Skerlj, 1937).

In addition to the visual acuity advantage enjoyed by dark-eyed people, it also appears that dark-eyed groups have a lower incidence of color blindness. The proportion of the male population with color blindness is about 8% for Northern Europeans and American whites, about 5% for Chinese and Japanese, and less than 5% for American blacks, Mexicans and American Indians (Specter, 1956).

Two cautions about the above figures should be emphasized. One is that they refer only to males. The incidence of color blindness is much lower in females. The other point to be stressed is that the figures given are only approximations based on one or more samples taken from those populations. Taking the figures as a whole, however the lighter-eyed groups tend to have more color blindness than the darker-eyed groups.

We also know that many people who are color blind also have low visual acuity, increased photophobia (i.e., excessive sensitivity to light) and a visual spectrum which is shortened at the red end (Walls, 1963). Those are all characteristics associated with light eyes.

The disadvantage at the red end of the spectrum is offset by greater sensitivity at the other end of the spectrum and it is clear that eye pigmentation is an important element in determining these differential acuities and sensitivities. That conclusion is consistent with Wall's (1963) statement that blue-blindness is related to "excessive absorption of short-wave light in an abnormally rich macular pigmentation . . ." (p. 99). Red blindness, then is more of a problem for light-eyed people and blue blindness is more of a problem for dark-eyed people.

The findings based on people with defective vision can be supplemented by some recent findings based on studies of people with normal vision. Psychologists in the area of perception have been interested in visual illusions. Some persons are more susceptible to these illusions than others. One popular illusion which has been studied is the Mueller-Lyer illusion. The figures used to study this illusion are shown in figure 8-1:

The middle line is the same length in both figures but there is a tendency to see the line in the top figure as longer. Various sophisticated techniques have been developed to measure the degree to which a person is susceptible to the illusion. When people from different ethnic groups have been tested, non-Caucasian groups have been found to be less susceptible to the illusion than are Caucasians. There have also been studies that indicate that as children get older they become less susceptible to the illusion. Pollack and Silvar (1967) noted that both these findings might be explained in terms of amount of ocular pigmentation. The non-Caucasian groups studied were more darkly pigmented in general than Caucasians. The eyes of children become more darkly pigmented with age. Thus, they hypothesized that susceptibility to the illusion would be inversely related to amount of eye pigmentation. To test the hypothesis they studied 35 male children for whom amount of optical pigmentation had been determined by means of an opthalmoscope (Silvar and Pollack, 1967). Each child was tested for susceptibility to the illusion and scores on that variable related to the eye darkness designation.

The results indicated clearly that those children with the more deeply pigmented eyes were less susceptible to the illusion. There was some overlap in susceptibility between black children and white children which was related to the fact that one black child had light-eyes and several white children had very dark eyes. The authors concluded, "Thus, optical pigmentation rather than race membership appears to be the more important variable affecting sensitivity to the illusion" (Pollack and Silvar, 1967, p. 84).

Figure 8-1

The Mueller-Lyer Illusion

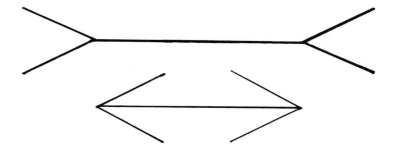

John Berry, A Canadian researcher, reports that whereas some subsequent studies have failed to find the phenomenon reported by Pollack and Silvar, he has reanalyzed some old data which provides strong support for their conclusion (Berry, 1971). Berry had earlier, as part of another study, tested ten groups of people for susceptibility to the Mueller-Lyer illusion. The groups were chosen to be culturally diverse and included Europeans, Eskimoes, Africans and natives of Australia and New Guinea. The ten groups were ranked on the author's estimate of skin pigmentation "as an index of retinal pigmentation". The total sample included data on almost 500 individuals from the ten groups. The group rank on pigmentation was highly related (Kendall's Tau coefficient of .82) to group rank on susceptibility to the illusion. The darker groups were less susceptible. Berry concluded, "The only conclusion possible from these data is that pigmentation is the best predictor of Mueller-Lyer susceptibility across ecological and cultural settings, even when confounding influences of theoretically competing variables are statistically removed" (p. 196).

As mentioned before, competing visual functions are such that an advantage in some areas of functioning will carry with it a reciprocal disadvantage in other areas. Such is the case here. People who are more successful in judging illusions such as the Mueller-Lyer are less successful in making discriminations that require sensitivity to visual contrast. Pollack (1963) found a significant negative correlation between illusion magnitude and "contour detectability threshold". So the dark-eyed person's advantage in one area of perceptual accuracy is offset by a disadvantage in another area of perceptual accuracy.

The differential advantage of disadvantage of eye pigmentation on these tasks is further complicated by the effects of pigmentation on color perception. Pollack and Silvar (1967) in their study of illusions used blue light because ". . . it was felt that exposure of a Mueller-Lyer figure in blue illumination would maximize differences between light and dark pigmented *Ss* by reducing the intensity of the illusion figure's contours in deeply pigmented *Ss* (p. 83). They also noted work done in Egypt by Ishak (1952a, b), ". . . who found that Egyptians exhibited markedly deeper macular pigmentation than Europeans and that this increased density was associated with higher luminosity thresholds in the shortwave region of the visible spectrum" (Pollack and Silvar, 1967, p. 83). This would suggest that the colors involved in illumination or materials used could increase or decrease a person's accuracy on a particular perceptual task as a function of his eye pigmentation.

That fact is important, for it may have practical implications in a number of areas—especially education. The point is well illustrated by a study (Jahoda, 1971) done with university students from Scotland and university students from the African country of Malawi. Jahoda was aware that African students tend to do poorly on some tasks that involve spatial relations and he reasoned that this problem might be more or less serious depending on the color of the materials used. The way he tested this idea was very straightforward. He used only students who had taken geography courses. He gave them a map-reading test that required attention to contours indicated by changes in shading. Half the items involved yellow/red land shadings and half involved purple/blue water shadings. The Scotish students did about equally well on the two types of items. They were slightly, but not significantly, better on the purple/blue water items. For the African students, on the other hand, the colors involved were very important. They did significantly better on the red/yellow land shadings than they did on the purple/blue water shadings.

Jahoda also found, consistent with his expectation, that African students were more susceptible to the Mueller-Lyer illusion when red figures were used than when blue figures were used.

He summarized the significance of his findings as follows.

The results relating to M-L susceptibility, confirming Pollack's hypothesis are mainly of theoretical interest. The extension of the same principle as applying to space perception, which may contribute towards an understanding of the difficulties of darkly pigmented people in this sphere, has potential practical implications. It suggests that careful attention should be paid in the preparation of instructional materials to the colors employed. Whether or not this would substantially improve spatial-perceptual performance remains a question to be explored by further research (p. 206).

Indeed, much research is yet to be done and it is important that it be done if we are to have knowledge useful in tailoring educational materials to the student's strengths rather than to his weaknesses. The same consideration should be given to the possible importance of different colors in non-educational settings—such as on the job. Unfortunately, many behavioral scientists are reluctant to study group differences because of the assumption that even if differences do exist it is better that we not know about them. The feeling here seems to be that the only possible use of such knowledge would be to justify social, cultural discrimination against one group by another. The Jahoda findings may lead to a reconsideration of that assumption.

Before concluding the topic of eye color and perception there is an additional observation to be reported. We have noted that dark eyes facilitate visual acuity and perception of bright colors, but hinder contour detectibility and other tasks involving spatial relations. Since differences in eye pigmentation are genetically determined, this difference in perception is probably genetic in origin. Consistent with that conclusion are statements made by Hess (1970):

A frankly biological approach to human behavior may be found in genetics. This is a very young field that promises to become extremely fertile, particularly when the ethological premise of the biological bases of behavior is accepted without racist handicaps. In my own unpublished research I have found a strong and highly consistent correlation between eye color and modes of perception and personality. Blue eyes were associated with form dominance in perception and a scientific attitude, whereas brown eyes were associated with color dominance in perception and a nonscientific attitude. The genetic bases of this obviously could be studied (p. 31).

In addition to the Hess observation, Markle (1972) found a relationship between eye color and form vs. color perception in responses to Rorschach ink-blot materials. The blots, which included bright colors, could be responded to in terms of color or form. The participants in the study were all male college students. Relative to each other, brown-eyed students responded more to the color of the blots and the blue-eyed students responded more to the form or shapes of the blots.

Response to color versus response to form is used by psychologists as an indication of personality. People who respond more to color are considered interpersonally spontaneous and emotionally reactive. People who respond more to form are considered slow, hesitant and emotionally controlled.

The findings reviewed here, taken together, again present a consistent picture. Eye color or eye pigmentation is found once more to be not a meaningless characteristic as most people suppose, but an important determinant of how the individual perceives his world and responds to it.

Differences in perception may be the underlying cause for some of the other differences between light-eyed and dark-eyed organisms covered in earlier chapters. However, I do not think that all behavioral differences can be traced to differences in visual perception. As already noted the eye serves photoreceptor functions other than vision. These functions are mediated by control centers in the brain that regulate endocrine functioning. We will review some of what is known about these functions in order to see how they, too, may help to explain our eye color findings.

References

Bassin, R. and Skerlj, B. Augenfehler and augenfarbe (Eye defects and eye color) *Klinische Monatsblatter Augenheilkunde,* 1937, *98,* 314-324.

Berry, J. W. Mueller-Lyer susceptibility: Culture, ecology or race? *International Journal of Psychology,* 1971, *6,* 193-197.

Brown, L. and Amadon, D. *Eagles, hawks and falcons of the world.* New York: McGraw-Hill Book Company, 1968.

Dreger, R. M. and Miller, K. S. Comparative psychological studies of Negroes and whites in the United States: 1959-1965. *Psychological Bulletin Monograph Supplement* 1968, *70,* no. 3, Part 2, 1-58.

Droscher, V. B. *The magic of the senses.* New York: E. P. Dutton, 1969.

Falls, H. F. Albinism *Transactions of the American Academy of Ophthalmology,* 1953, *57,* 324.

Gilliard, E. T. *Birds of paradise and bower birds.* Garden City New York: The Natural History Press, 1969.

Gordon, M. S. *Animal function: Principles and adaptations.* New York: McMillan, 1968.

Grant, V. *Flowers.* Encyclopedia Americana, 1959a, *11,* 390b-395.

Grant, V. *Pollination.* Encyclopedia Americana 1959b, *22,* 320-323b.

Hess, E. H. Ethology and developmental psychology in P. H. Mussen (ed.) *Carmichael's Manual of Child Psychology.* Vol. 1, New York: John Wiley and Sons, 1970, 1-38.

Iredale, T. *Birds of paradise and bower birds.* Melbourne: Georgian House, 1950.

Ishak, I.G.H. The photopic luminosity curve for a group of fifteen Egyptian trichromats. *Journal of the Optical Society of America,* 1952a, *42,* 529-534.

Ishak, I.G.H. The spectral chromaticity coordinates for one British and eight Egyptian trichromats, *Journal of the Optical Society of America,* 1952b, *42,* 534-539.

Jahoda, G. Retinal pigmentation, illusion susceptibility and space perception. *International Journal of Psychology,* 1971, *6,* 199-208.

Karpinas, B. D. Racial differences in visual acuity. *Public Health Report,* 1960, *75,* 1045-1050.

Markle, A. Effects of eye color and temporal limitations on self-paced and reactive behavior. Unpublished doctoral dissertation, Georgia State University, 1972.

Pearson, T. G. (Ed.) *Birds of America.* Garden City, New York: Garden City Publishing Company, 1936.

Pollack, R. H. Contour detectability threshold as a function of chronological age. *Perceptual and Motor Skills,* 1963, *17,* 411-417.

Pollack, R. H. and Silvar, S. D. Magnitude of the Mueller-Lyer illusion in children as a function of pigmentation of the Fundus oculi. *Psychonomic Science,* 1967, *8,* 83-84.

Silvar, S. D. and Pollack, R. H. Racial differences in pigmentation of the Fundus oculi. *Psychonomic Science,* 1967, *7,* 159-160.

Specter, W. S. (ed.) *Handbook of biological data.* Philadelphia: W. B. Saunders Company, 1956.

Walls, G. L. *The vertebrate eye and its adaptive radiation.* New York: Hafner Publishing Company, 1963.

Chapter 9

Light and its Psychobiological Effects

We have seen that amount of pigmentation influences to some degree how much light and which particular wavelengths of light will enter the eye. It was mentioned that the eye is involved in important non-visual functions. It is possible, and I think very probable, that the mediation of reactive and non-reactive behavior, as well as other differences correlated with eye color, is linked to light-related physiological functions. In this chapter, then, we shall consider some of the psychobiological effects of light.

Some Examples of Effects of Light

The Varying Hare was given its name because in the summer its coat is brown but in the winter it is white. That is a convenient adaptation for the hare since it helps the hare blend into the environment and escape from predators. The Varying Hare's white winter coat means survival when snow is on the ground, but it would be very disadvantageous in the summertime. The opposite, of course, is true for the brown summer coat. One might assume that the change in coat color from brown to white and back again is triggered by a change in temperature. That is not the case, however. Rule (1967) states:

> Its eye is in effect a photoelectric cell that is activated by the number of daylight hours. The eye picks up less light in the short daylight hours of autumn, thus allowing the pituitary gland at the base of the animal's brain to become inactive. This gland controls the amount of pigment that goes into each hair of the hare's body. When the pituitary gland is inactive, no pigment is formed in the hair and the new winter growth of hair is white (p. 28).

In the spring, the process is reversed with longer periods of light activating the pituitary gland which triggers melanization of the hare

97

hair. Controlled studies have shown that the coat color can be changed back and forth by artificially controlling the amount of light to which the hare is exposed. So the hair color, mediated by changes in an endocrine gland, varies naturally on a yearly cycle which is controlled by light. We have already noted that some animals (e.g., certain lizards) undergo color changes on a daily cycle.

We will have more to say about how light influences cyclic behavior. We will also give more consideration to endocrine glands and how they are influenced by light. But first let us look at some illustrations of the diversity of light's effects.

An interesting story is told by Hollwich (1964) of an Austrian deer hunter, Von Schumacher. Von Schumacher noticed that the number of roe horns awarded prizes at the Tyrol hunting exhibition differed greatly from one year to another. Prizes were given only if the horns, which are secondary sexual characteristics, reached a certain level of development. A very objective scoring system was used to determine development, but even so, the number of prizes given varied greatly. This was true even though the number of hunters tended to be about the same each year. One year there was only one prize given; another year 32 prizes were given. Von Schumacher thought that something other than chance was operating. He kept searching until he found an explanation. The clue which he finally found was in the weather records that indicated the amount of cloudiness in the area during each season. Roe horn development for any given season was positively correlated with the total hours of sunshine. We have here a case of light influencing a physical characteristic indirectly related to reproduction.

Zacharias and Wurtman (1964) became interested in the effects of light on rate of sexual maturation in blind girls. They were aware that in birds and other animals seasonal changes in sexual behavior were related to light. They had also seen an anecdotal report that the advent of springtime elicited behavior in Eskimoe which had been given the descriptive label, *furor sexualis*. They reasoned that if light really is directly related to the human reproductive system, blind girls, who have experienced less light input than nonblind girls might differ from them in rate of sexual maturation. The data collected from each group was the age at menarche. Their analysis showed that there was a difference. The blind girls matured sexually at an earlier age than the normally-sighted girls. Furthermore, those girls that were totally blind matured faster than those that were partially blind.

There are other reasons to believe that a less than normal light input affects physiological functioning. Blind people differ from nonblind in water excretion, water balance, carbohydrate balance, and insulin balance (Hollwich, 1964). Changes may also occur from a greater than normal input of light. Hague (1964) reports an interesting case involving a reunion of soldiers who years before had been exposed to an extremely high light intensity. When they compared notes they found that an unusual number of them were either sterile or had deformed children. It has also been reported (Hollwich, 1964) that women who move to the tropics are temporarily unfruitful. This is possibly caused by changes in sex glands resulting from the change in environmental light.

The examples we have considered show that among other things, light has an influence on sexual development and functioning. Such information would come as no surprise to chicken farmers. They systematically manipulate light and dark cycles in their hen houses in order to influence rate of maturation and egg production. Hens reared in permanent light start laying late. However, they produce more and heavier eggs (Jochle, 1964).

Continuing, the next sample involves the effect of light on another area of physical functioning or rather dysfunctioning, obesity. There is some indication that persons deprived of light are more likely to be obese. This relationship was apparently known by the ancient Egyptians. A study of Egyptian drawings (Fuchs, 1964) revealed that figures depicting blind people showed them as obese. Jochle (1964) reports that similar observations about light were made in 1799 by a German, Johann Ebermaier, who wrote a book based on observations of the blind, prisoners, and others deprived of light. "Worth mentioning is his feeling that obesity in blinded animals and in prisoners kept lightless was not only effected by inactivity, but resulted from a disturbance in metabolism caused by deprivation of light" (Jochle, 1964, p. 89). If modern physiologists are able to confirm those observations and gain some insight into the metabolic processes involved, perhaps, one day we can stop all our dieting and spend our money on special weight-control lamps.

Lest that possibility sound too fanciful, let us consider now some examples of psychobiological changes that have resulted from artificial changes in light source. John Ott, the director of the Environmental Health and Light Research Institute has been very concerned about the effects of different kinds of artificial lighting on health. Ott first became interested in these effects because of some accidental findings that grew out of problems he was having as a photographer of plant growth (Ott, 1964). He was making time-lapse photographs to be used for such things as showing a flower develop and bloom. He found that, unless he was very careful, the lighting involved in his photographic procedures would disrupt the natural growth process. He noted that the effects of artificial changes in light were wavelength dependent.

On one job he changed the wavelengths of light reaching part of a plant and lost several months work as a result. He had been commissioned to get time-lapse color photographs of an apple ripening for a Walt Disney picture. To do this he set up his camera to periodically take pictures of a particular limb of an apple tree. By showing the frames in rapid sequence as a moving picture a year's growth could be reduced to minutes and an apple would grow and ripen "before your very eyes." To protect his equipment he covered the limb with a protective glass covering. Unfortunately, the glass he selected filtered out some wavelengths of natural light. For a while there was no problem. All through the spring and summer the apples developed nicely and were duly filmed. The problem was noted only when most of the apples on the tree stopped growing and ripened into a nice red color. Those apples being photographed stayed green and kept growing to twice the size of the others. Only by changing the glass to one which was more neutral was he able to film the natural sequence without disrupting it.

This and similar experiences led Ott to begin systematically gathering data on the effects of light. He has since reported (Ott, 1968) numerous other examples of effects that followed from artificial manipulation of light wavelength. In addition to his formal studies, he has collected some very interesting anecdotal evidence. One example involved unexpected benefits that followed from a change in artificial lighting. One hotel restaurant in Chicago, for ornamental and decorative purposes, used lights and filters (black ultraviolet) that make the light in the restaurant mostly short wave light. This lighting has been used for seventeen years and the management of the hotel has been surprised to find that employees in the restaurant differ from their other employees in having much better health records with fewer cases of colds, flu, and virus infections.

Another example involved some unexpected difficulties that followed from a change in artificial lighting. Again for decorative purposes, a radio station in Florida was equipped with pink (long wavelength) lighting. It was noted following the change that workers had become more emotional and dissatisfied and were frequently irritated with one another. Performance of announcers deteriorated with a corresponding increase in errors until someone finally made the connection between the change in behavior and the change in lighting. The pink lights were removed and things returned to normal.

These examples indicate that light influences our lives in a wide range of biological and behavioral areas. Pigmentation, especially of the eye, determines how much light and what wavelengths of light enter the body. Knowledge about the physiological effects of pigmentation may help us understand some of the psychobiological differences between light-eyed and dark-eyed organisms. Such knowledge may also be a helpful source of additional hypotheses about eye color differences.

Light and the Endocrine System

In the examples that were given, a number of bodily functions were seen as related to light. These various functions are apparently involved with the intricate interworkings of the various endocrine glands. These glands include the pituitary, pineal, thyroid, parathyroid, thymus, gonads, and adrenals. Each gland secretes hormones that circulate in the blood and control specific functions. Each year endocrinologists learn more about these functions, but there is still much that is not known. Part of the problem in understanding the endocrine system arises from the facts that (1) the same hormone may have more than one effect (2) the same effect may be caused by more than one hormone, and (3) hormones from one gland influence the output of other glands. Furthermore the homeostatic feedback system is such that an increase in a particular hormone sets in motion a process which operates to decrease the production of that and perhaps other hormones.

These glands, individually and in concert, control, by means of hormones, such things as growth, pigmentation, sexual behavior, blood sugar level, and biological response to stress. The pituitary gland in the middle of the brain is referred to as the "master gland" because it greatly influences the workings of the other glands. The pituitary, in turn is influenced by a nearby organ, the hypothalamus. Together, the hypothalamus and pituitary are considered the regulators of endocrine functions. Since these two organs are directly and indirectly responsive to light, it is partly through them that light has its effects on endocrine functions. These two organs, together with the retina in the eye, play a major role in "photo-neuro-endocrine biology" (Hague, 1964). The pineal gland, which will be

101

discussed in the next chapter, is also a part of that system.

Wavelength of Light and Endocrine Response

As we have discussed previously, one effect of differences in eye pigmentation is to influence the relative amounts of different wavelengths of light that enter the eye. Melanin blocks short wave (blue) light much more effectively than it blocks long wave (red) light. Therefore, heavy pigmentation (in any part of the body surface, including the eye) results in mostly red light getting through. Decreased pigmentation has the effect of letting in both red and blue light. So, when we consider the relative proportions of red and blue light that enter the eye we can say that darkly-pigmented eyes are more receptive to red light and lightly-pigmented eyes are more receptive to blue light. Could that make any difference in endocrine functioning? Apparently so.

Since differently pigmented irises serve as filters that let in different wavelengths of light, we can use as a type of simulation, studies in which organisms have been exposed to light which has been artificially filtered such that it is more red or blue than is natural (white) light.

The example of the radio station that changed the light to make it pink suggests that biological changes could be involved. That is, red light may make us more sensitive or autonomically responsive than other colors of light.

There have been controlled studies that demonstrated directly that different colored lights have had effects on endocrine glands or on behaviors that clearly resulted from hormonal changes. Hollwich (1964) reported that when young male ducks were continuously exposed to orange or red light, gonad size increased to a level much above normal. Benoit (1964) has demonstrated that it is only reddish light which causes the gonads to grow. Short wavelengths of light do not have that effect.

Walls (1963) reported on a study done with mud-minnows and shiners by Reeves (1919) the results of which demonstrate the activating effect of red light. Reeves exposed the fish to a red light and found that respiration rate more than doubled. "In this experiment it was perfectly clear that the response was to redness as such, since the respiration rose with an increase of brightness, but rose still higher when that brightness was somewhat reduced by a filter which introduced hue" (Walls, 1963, p. 486).

Hernandez (1940) did a study that demonstrated that humans,

also, have different autonomic responses to different colored lights. His measure of autonomic response was the galvanic skin response (GSR). The GSR measures changes in skin electrical conductance that result from slight changes in the sympathetic nervous system. The GSR is a very sensitive measure of autonomic reactivity and is one of the indicators used in lie detectors. Hernandez flashed different colored lights and measured the GSR to each. The greatest response was to red and the least response was to blue.

The organisms we have considered, so far, are all rather complex. Actually even one-celled animals can and do respond differently to different wavelengths of light. Droscher (1969, p. 52) reports that the unicellular creature, *amoeba proteus,* can discriminate red and blue. When exposed to red light its movements accelerate; when exposed to blue light its movements slow down.

Finally, Ott (1968) reported that studies of animals indicate that different wavelengths of light have different effects on insulin balance, toxicity of drugs, tooth decay, and sex ratio of offspring. All of these effects are probably related to differences in endocrine functioning.

Direct Effects of Light on the Endocrine System

The eye, as a specialized photoreceptor, is a very important adaptation, but the body is not totally dependent on the eye for light input. Benoit (1964) demonstrated that endocrine effects could be obtained by removing the eyes of the organism (male ducks) and shining a light through a glass rod directly on the hypothalamus. When this was done testicular growth responded to changes in the light in a manner consistent with what results from normal changes in environmental light. Another study (Lisk and Kannwischer, 1964) used the same technique to shine light directly on the hypothalamus of female rats. The weight of the pituitary gland and the ovaries increased as a result of this light.

Given that the hypothalamus, and perhaps other brain centers also, can respond directly to light, the question becomes whether, in the normal course of life, light penetrates to the brain. The answer is, yes. Light does reach the brain—through the eye socket (over and above that mediated by the retina) and even through the skull. This is a normal part of input to the intact animal and it also can serve as a back-up system for the eyes. Again, however, not all wavelengths penetrate equally well. Typically, it is only the long wavelengths that reach the hypothalamus directly (Benoit 1964).

The idea that light penetrates the body and that more long wave than short wave light gets through may at first seem a bit strange or hard to believe. Menaker (1972), in an article on nonvisual light reception, has suggested a way to satisfy such doubts in a very simple manner:

> In a completely dark room place an ordinary flashlight against your palm, switch it on and look at the back of your hand. This should convince you that light does penetrate living tissue, and furthermore that it is the long wavelengths that penetrate best. In effect the human hand, although it is almost an inch thick, is a moderately transparent red filter (p. 29).

As mentioned before, the ability of the endocrine control centers to respond directly to penetrating light may at times serve as a back-up system for the eyes, but it may also be a part of normal functioning. Recently, researchers from the University of Tokyo and the University of California at Davis (Homma, Wilson, and Siopes, 1972) have jointly advanced a hypothesis, based on findings with birds, that ocular and encephalic photoreception serve different purposes. They found some evidence to indicate that light received directly by the brain was important in turning the reproductive system on and that light mediated by the eyes was more important in turning the reproductive system off. Since the eyes can apparently receive short wavelengths of light better than can the brain directly, it would seem a reasonable possibility that short wavelength light serves to inhibit endocrine functioning. Since reception of short wavelength light is more dependent on the eye than is long wavelength light, blindness would perhaps result in some endocrine glands being more active than they would be otherwise. Dark pigmentation of the eye might have a similar effect since it, too, would serve to let in relatively less short wavelength light.

Light and Biological Clocks

The endocrine functions of a living organism are never static. The hormonal production of glands ebbs and flows. Body chemistry in every organ and every cell is constantly changing. That may sound chaotic but it is not. The changes are not random, but rhythmic and coordinated one with another. The rhythms vary in length from a fraction of a second to a year or more and probably evolved in the species, like everything else, by providing superior survival value. Take for example, the yearly rhythm of sexual functioning found in many birds. The increase in sex hormones in springtime causes the birds to mate at precisely that time of year when warm weather and increased food supply gives them the best chance to successfully rear offspring. Operating on that yearly cycle has obvious survival value for the species and only those birds that did operate on that cycle survived. The physiological functioning of the bird has been selected to meet environmental demands. Because the environmental changes are on a yearly cycle, the hormonal cycle is yearly also.

Likewise, the environmental demands for most animals change on a 24-hour, daily cycle. What is optimal survival behavior during the day may not be optimal survival behavior at night. Since behavior and physiological functioning are closely related, it is not surprising that many bodily rhythms are circadian (about 24 hours in length).

In addition to yearly and daily cycles there are also, for some organisms, monthly physiological cycles. The sexual activities of many sea creatures for instance, are on a monthly cycle. This cycle is related to changes in the moon which presents yet another natural light cycle. Perhaps the length of the human menstrual cycle also originated as a cycle coordinated with changes of the moon.

Most biological rhythms appear to be yearly, daily, or monthly in length, or, if shorter than a day, of a length which divides equally into 24 hours. This suggests that rhythmicity or timing is yet another biological function related to light reception. Although experimental work has been done in this area for more than a hundred years, it is only recently that the subject of "biological clocks" has become an extremely popular area of research. Ritchie Ward in his 1971 book, *The Living Clocks* presents a very readable summary of this fascinating area.

Biological clocks are not unique to the animal kingdom; even plants are capable of doing things which require timing. For instance, the leaves of some plants turn up during the day and droop at night. This is done on a 24-hour cycle and the plant responds on schedule even if placed in a closet where it is not exposed to light. The plant's clock, it is true, will tend to run fast or slow after a while, but the same thing happens to the biological clocks of animals when they are exposed to constant light or constant darkness. However, when the plant or animal with a slow or fast rhythm is exposed briefly to natural light it quickly reverts to a cycle that is exactly 24 hours long.

Several studies have shown that cycles shorter than a day are also coordinated into the 24-hour cycle. Richter (1971) found that Norway rats are active for exactly 12 hours a day and inactive for 12 hours. He reasoned that the rats had evolved originally in the tropics where light and dark periods are exactly 12 hours long. The Pileated Tinamou, a bird that lives in Panama, sings at 3-hour intervals, day and night. This 3-hour cycle breaks the 24-hour day into 8 equal parts.

Studies have found that animals have difficulty responding to time periods that are not consistent with natural units of time. A bee can learn to come for food at the same time each day, but it cannot learn to come every 19 hours (Ward, 1971, p. 104). Recently it has been demonstrated that hamsters do not respond to changes in the light/dark cycle if the cycle is manipulated in such a way that it is inconsistent with a 24-hour clock. With cycles of 24 or 48 hours duration, the hamster's testes increased or decreased in size according to the length of the light period relative to the dark period as it would in different times of the year. However with cycles of 36 or 60 hours duration the changes in testes size did not occur (Elliot, Stetson and Menaker, 1972).

Some information about biological cycles can be put to practical use. For instance, flies have been found to be several times more susceptible to insecticide in the afternoon than in the morning. Likewise diabetic patients are known to be more sensitive to insulin at night than in the daytime. It is also reported that many illnesses and emotional problems are periodic in their manifestations. As more is learned about biological rhythms this knowledge will be more and more useful in medical research and medical practice.

For those who have specialized in this general area of research, one of the chief questions has been whether biological rhythms are dependent on or independent of *present* environmental changes. The evidence indicates that almost all biological rhythms are endogenous (i.e., the length of the cycle is an inherited characteristic not dependent on the length of *present* environmental cycles). Across many generations, cycles evolved which were consistent with environmental light cycles. In that way the rhythms became a part of the organism which can maintain cyclic functions independent of present environmental light input.

Environmental light, however, is important in synchronizing the various endogenous rhythms. To do this job, one endocrine gland, the pineal, differs from the others in not functioning according to an endogenous rhythm. Its production of hormones is externally controlled by changes in light input to the eyes (Wurtman and Axelrod, 1968). The pineal apparently serves as a special biological clock which coordinates various rhythms in the body. This uniqueness gives the pineal gland a special role in regulating endocrine functions. Because of its special relationship to the eyes and to endocrine regulation, the pineal gland may very well be an important link in the relationship between eye color and psychobiological variables. That possibility makes it worthwhile for us to give it some special consideration. The next chapter will report some of what is now known about this little gland which at one time was called the pineal "body" because many physiologists thought that it was not a gland and in fact had no function at all.

References

Benoit, J. The role of the eye and of the hypothalamus in the photo-stimulation of gonads in the duck. *Annals of the New York Academy of Sciences,* 1964, *117,* 204-215.

Droscher, V. B. *The magic of the senses.* New York: E. P. Dutton, 1969.

Elliot, J. A., Stetson, M. H., and Menaker, M. Regulations of testis function in golden hamsters: A circadian clock measures photo-periodic time. *Science,* 1972, *78,* 771-773.

Fuchs, J. Physical alternations which occur in the blind and are illustrated on ancient Egyptian works of art. *Annals of the York Academy of Sciences,* 1964, *117,* 618-623.

Hague, E. B. Opening remarks to a conference on "Photo-neuro-endocrine effects in circadian systems, with particular reference to the eye." *Annals of the New York Academy of Sciences,* 1964, *117,* 5-12.

Hernandez, J. E. The perception of yellow light through red and green binocular stimulation as determined by the conditioned galvanic response, *Journal of Experimental Psychology,* 1940, *26,* 337-344.

Hollwich, F. The influence of light via the eyes on animals and man. *Annals of the New York Academy of Sciences,* 1964, *117,* 105-131.

Homma, K., Wilson, W. O., and Siopes, T. D. Eyes have a role in photo-periodic control of sexual activity of coturnix. *Science,* 1972, *178,* 421-423.

Jochle, W. Trends in photophysiologic concepts. *Annals of the New York Academy of Sciences,* 1964, *117,* 88-104

Lisk, R. D. and Kannwischer, L. R. Light: evidence for its direct effect on hypothalamic neurons. *Science,* 1964, *146,* 272-273.

Menaker, M. Nonvisual light reception. *Scientific American,* 1972, *226,* 22-29.

Ott, J. N. Some responses of plants and animals to variations in wavelengths of light energy. *Annals of the New York Academy of Sciences,* 1964, *117,* 624-635.

Ott, J. N. The influence of light on the retinal hypothalamic endocrine system. *Annals of Denistry,* 1968, *27,* 10-16.

Reeves, C. D. Discrimination of light of different wavelengths by fish. *Behavior Monographs,* 1919, *4,* No. 3, p. 106.

Richter, C. P. Inborn nature of the rat's 24-hour clock. *Journal of Comparative and Physiological Psychology,* 1971, *75,* 1-4.

Rule, L. L. *Pictorial Guide to the Mammals of North America.* New York: Thomas Y. Crowell Company, 1967.

Walls, G. L. *The vertebrate eye.* New York: Hafner Publishing Company, 1963.

Ward, R. R. *The living clocks.* New York: Alfred A. Knopf, 1971.

Wurtman, R. J. and Axelrod, J. The pineal gland. *Scientific American,* 1968, *213,* 50-60.

Zacharias, L. and Wurtman, R. J. Blindness: Its relation to age of menarche. *Science,* 1964, *144,* 1154-1155.

Chapter 10

The Third Eye and the Pineal Gland

It was pointed out in earlier chapters that the eye has two somewhat competing functions, vision and photosensitivity. Decreased eye pigmentation decreases visual acuity, but it makes the eye a more sensitive receptor of light. In the last chapter, it was pointed out that light has an effect on the endocrine system in the body. This effect is mediated by the actions of the endocrine regulator, the hypothalamus, on the "master gland", the pituitary. It is now known that these organs are to some degree regulated, in turn, by the pineal gland which has the capacity to inhibit their hormonal secretions. The pineal gland is the organ in the body perhaps most dependent on nonvisual light input from the eye.

This input is first obtained by the retina. It is not known what type of cells are involved nor the degree to which they are differentially sensitive to different wavelengths (Wurtman, 1969). That the wavelength of the light does make a difference is indicated by research done in Yugoslavia (Miline, 1971) which revealed that in rabbits exposed to red light, the pineal decreased in size and function. It has also been suggested that blue light may have a special facilitative effect on the functioning of the pineal (Wurtman, 1971). So, both on the basis of absolute sensitivity and sensitivity to different wavelengths, eye pigmentation would seem to have a possible influence on the functioning of the pineal gland.

The eye transmits some information by means of direct nerve connections with the pineal gland. Other information from the retina, that having to do with vision, is transmitted to the brain via the primary optic tracts. A separate system, the inferior accessory optic tracts, transmits nonvisual information from the eyes, through the hypothalamus to the pineal gland. The hypothalamus may receive some light information directly from these nerves as well as indirectly from the pineal (Wurtman, 1971b).

The pineal gland, because of its unique nature—it is the only

110

unpaired organ in the brain—was considered by ancient peoples to have unique and important functions. According to ancient Indian literature the pineal functions as an organ of clairvoyance and meditation allowing man to remember his past lives. Even today, there are religious cults that consider the pineal as the source of all insight and wisdom. The ancient Greeks thought that the pineal controlled all thoughts, and Galen writing in the second century A. D. saw the pineal's function as coordination of sensory and motor activity. Descartes, in the 17th century, considered the pineal the "seat of the soul" (Kappers, 1966).

When modern endocrinologists came along to determine scientifically what function each of the glands serve in the body, they thought for a time that the mammalian pineal was, perhaps, only a vestige of an earlier organ with little or no present function. For that reason, it was referred to as the pineal body rather than the pineal gland. Now we know that it is indeed a gland, and a very important one. Before reviewing some of what is now known about the functions of the pineal in birds and mammals, it will be well to consider its funtion in lower animals.

The Pineal in Lower Animals

The pineal is well developed in many lower animals and in some it has its own eye. This "third eye" or "pineal eye" was mentioned earlier as one adaptation which gives the organism greater sensitivity to light. Apparently, having an extra hole in the skull had some serious disadvantages for higher organisms. They have evolved a system whereby the pineal does not require an extra eye, but gets information from the two lateral eyes. From time to time, humans and other mammals are born with a "third eye" in the middle of the forehead but this is the result of a severe genetic malfunction and individuals with this condition have not lived very long (Droscher, 1969, p. 49).

The pineal eye is unusually well developed in the New Zealand lizard, the tuatara, which is the last survivor of a family of reptiles that flourished millions of years ago. The tuatara is noted for its capacity to slow down or inhibit vital processes. It has been observed, for instance, to wait an hour between breaths. It is also noted for very slow maturation and does not become sexually mature until it is about 20 years old (Bellairs and Carrington, 1966).

The pineal perhaps plays some role in the tuatara's slow growth and sluggish activity. This would be consistent with the fact that lizards in which the pineal has been removed become abnormally restless and expose themselves to the sun more than is common in normal lizards.

Also, in lizards the pineal mediates changes in skin color that serve to camouflage the lizard by making it the same color as its surroundings. In some fish, too, the pineal allows the fish to change colors to match different backgrounds. To do this, the pineal system has to respond differently to different wavelengths of light.

Studies with lower animals (e.g., fishes and lizards) done by Dodt and his co-workers in Germany have shown that the pineal system responds to light by changing nervous impulses transmitted to the brain. The impulse frequency is increased in response to medium and long wavelengths and decreased in responses to short wavelengths. Long wavelengths, are said to stimulate an "excitatory component" and short wavelengths to stimulate an "inhibitory component" (Oksche, 1971).

The Pineal in Birds and Mammals

It is only very recently that much has been learned about the function of the pineal in birds and mammals. One reason that more was not learned earlier was that no one had been able to isolate and identify hormones secreted by the pineal. Finally, in 1958, Aaron Lerner and his co-workers at Yale University identified a hormone produced only by the pineal gland. They named it melatonin because of its blanching effect on melanin in frog skin. Subsequent studies with melatonin have shown that it has effects on many endocrine functions. In general, it seems to have its effect by inhibiting the actions of other glands.

Another reason that understanding of the pineal gland was slow in coming was that its effects are dependent on many interacting variables. Wurtman and Anton-Tay (1969) state, "The responses of the experimental animals to pineal compounds or their absence varies markedly with their age, sex, species and strain, and environment" (p. 507). Because of this complexity, it will be a long time before the pineal is fully understood, but studies have been pouring forth in the last few years and the pattern of functions is becoming clearer. In addition to serving a function in the synchronization of biological clocks, as mentioned in the last chapter, the pineal has other important functions.

Much evidence suggests that the pineal in lower animals is involved not only in the control of mobility, but especially in the control of adaptive pigmentation. For that reason, Lerner, when he discovered melatonin, expected it to have an effect on human skin pigmentation, but it did not. The pigment cells in human skin differ from those in lower animals and apparently once melanin is deposited in human skin it is not readily withdrawn as is the case in some lower animals. There is some connection, however, between the mammalian pineal and pigmentation since melatonin has been found effective in treating animals with disease that involves hyper-pigmentation (Wurtman, Axelrod and Kelly, 1968, p. 182). Also, melatonin has been found to inhibit the pituitary's secretion of melanocyte-stimulating-hormone (MSH) which plays a role in pig-mentation (Wurtman and Anton-Tay, 1969). So, the mammalian pineal seems to be able to inhibit pigmentation but less able to reverse pigmentation, at least in the human skin.

The first function of the pineal ever noted in humans had to do with the rate of sexual maturation. This was noticed because of spontaneous tumors in children which resulted in greatly increased or greatly decreased functioning of the pineal gland. If the tumor increased function the children were delayed in reaching puberty. If the tumor decreased the function of the pineal, puberty was reached at a very early age. Consistent with those observations, melatonin has been found to inhibit gonad growth and function in rats, quail, chickens and mice (Wurtman, Axelrod and Kelly, 1968, p. 154). Consistent with that, animals submitted to a pinealectomy (surgical removal of the pineal gland) reach puberty earlier and exhibit greater gonadal function (Moszkowska, Kordan and Ebels, 1971). This inhibition by the pineal of gonadal functioning is probably most important in those animals that vary in sexual functioning according to changes in environmental light (e.g., animals whose reproductive cycle is triggered and terminated by longer or shorter days as the season changes).

Another function of the pineal gland, that having to do with inhibition of activity, is the one potentially most relevant to reactive and nonreactive performance. Activity is greatly decreased when animals are injected with pineal extracts. Reiss and his co-workers at the Neuro-endocrine Research Unit, Willosbrook State School, New York, demonstrated that when animals (mice, rats and rabbits) were injected with a pineal extract there was a considerable reduction in activity. In pinealectomized animals (rats) on the other hand, there was a considerable increase in activity (Reiss, Davis, Sideman and Plichta, 1963).

Another study from the same laboratory (Reiss, Sideman and

113

Plichta, 1967) demonstrated this relationship between the pineal and activity in yet another way—one which would seem to be especially relevant to our findings on differences in speed between light-eyed and dark-eyed organisms. Reiss and his co-workers noted that even though specific rats were consistent in their activity from one day to the next there were great individual differences among the different rats. This was true even of litter mates. Based on observations of behavior in an activity—wheel it was possible to divide the rats into a "fast running" group and a "slow running" group. After it was clear which rats were in which groups they were all killed and histological studies were done of the pineal glands. A clear-cut difference was noted in the cell densities of pineal glands and adrenal glands for the two groups. The pineals of the slow running rats had high cell densities indicative of high hormonal production. On the other hand, they had less developed adrenal glands. The fast running rats had better developed adrenal glands and less developed pineal glands. These studies would seem to provide evidence for the idea that the pineal gland can operate to inhibit motor responses.

The same conclusion can be drawn from other studies as well. Research done in Italy (Martini, 1971) has involved testing the effect of melatonin on the learning of conditioned avoidance responses. The finding was that melatonin had no effect on the original learning but it did have a facilitative effect on extinction. That is, melatonin did not aid the animal in learning to respond but it did aid in learning *not to respond* after a response became unnecessary.

Light-eyed animals have a tendency to freeze in aversive situations. Drug research has shown that the tendency to freeze can be reduced by administering amphetamine (Winston, Lindzey and Conner, 1967). Melatonin, in turn, has been found to suppress the effectiveness of amphetamine (Wurtman, Axelrod and Kelly, 1968, p. 160). It would appear, then, that the pineal has an important role in mediating freezing behavior and probably other less extreme forms of inhibition of activity.

Studies with other drugs also indicate the same general pattern. Melatonin prolongs the sedative effects of phenobarbitone and increases the duration of sleep in bartituate-treated animals (Martini, 1971). Likewise, studies done by Nir (1971) in Israel have shown that pinealectomized rats do not die as quickly as normal rats when given a lethal dose of phenobarbitone. These findings indicate that the pineal can act as a suppressant to reduce the excitability of the central nervous system (Reiter, 1973).

Nir also found that the brains of pinealectomized rats differed

114

from normal rats in increased excitability and fewer slow brain waves. This is consistent with findings obtained with humans (Anton-Tay, 1971) which indicate that melatonin increases EEG synchronization and increases the amount of Alpha (slow) waves. These human subjects treated with melatonin also fell asleep easily and reported vivid dreams and unusual visual imagery.

Studies have now found that the pineal acts directly or indirectly to inhibit or moderate the functioning of most or all endocrine glands. This means that the pineal may have an effect on numerous bodily functions. As an example, melatonin has been demonstrated to be an inhibitor of contractions in smooth muscles in various parts of the body (Wurtman and Anton-Tay, 1969).

Research done in Romania (Milcu, Nanu-Ionescu and Milcu, 1971) has revealed that the pineal gland plays an important role in carbohydrate metabolism. Pinealectomized animals were found to exhibit higher blood sugar levels and other diabetes-like symptoms. This would seem to raise the possibility that a dysfunction of the pineal gland is involved in some types of diabetes and is perhaps involved in obesity as well.

There are other scattered findings, reviewed by Wurtman, Axelrod and Kelly (1968), that point to the possibility that greater knowledge of the pineal will lead to improvement in the treatment of various physical disorders. Pinealectomies in rats, for instance, result in systolic hypertension (high blood pressure) and certain subnormal immunological responses. The pineal gland appears, also, to be involved in the body's defenses against cancer. Studies have found a positive relationship between pineal function and the survival time of rodents infected with transplanted malignant tumors. Also, autopsies of persons who have died from malignancies have found the pineal gland enlarged, which would probably indicate that it had been overly active in response to the malignancy.

An early study (Altschulle, 1957) found that injection of pineal extract had a positive effect on the behavior of schizophrenic patients, but unfortunately the effect was very temporary. More recently, Anton-Tay (1971) has reported on studies done in Mexico which tested the effect of melatonin on epileptic patients and patients with Parkinson's disease. He reported EEG changes and mental changes in both types of patients, and in those patients with Parkinson's disease there was noted "a striking amelioration of all signs and symptoms of the syndrome" (Anton-Tay, 1971, p. 364). This research is still in the early experimental stages, but it may in the future have great practical importance.

One final study involving the pineal gland will be reported. These findings are especially relevant to what is known about eye color differences in behavior. It will be recalled that Tryon (1931) found that rats with light eyes were better at learning a maze than were rats with dark eyes. He attributed this to a tendency on the part of dark-eyed rats to be overly reactive in the maze situation. In 1964, Woolley and van der Hoeven, working with mice, demonstrated that administration of melatonin influenced maze performance. By administration of drugs, one of which, tyrosine, is a precursos of melanin pigmentation, they were able to produce a deficit in maze learning. They found, however, that if the mice were also administered melatonin the deficit did not occur. The pineal, then, is apparently involved in this ability which has already been shown to be related to eye darkness.

So, to conclude, although we cannot state with certainty that the pineal gland and its related organ, the hypothalamus, are the agents that mediate behavioral differences related to eye color, it seems a good guess, at this time, based on a great deal of inferential evidence.

Summary Statement to Part 3

In the last four chapters we have discussed anatomical and physiological facts that are probably related to the relationship between eye color and behavior. No effort has been made to specify exactly the physiological steps involved, but only to show that what is being found at the behavioral level is not implausible based on what is being found by contemporary research in the areas of perception and physiology. Demonstrating plausibility, I have learned, is required of anyone whose research produces findings that are new and unexpected.

In the next section, Part 4, we will consider some additional differences between light-eyed and dark-eyed people and light-eyed and dark-eyed animals. Relationships in these areas are more tentative and speculative than those already discussed. The data points are fewer and we must rely more on inference and generalizations from scattered facts. Even so, there are already enough findings in these areas to indicate that eye color research is almost certain to turn up surprising and important relationships in many areas of human and animal life.

References

Altschulle, M. D. Some effects of aqueous extracts of acetone-dried beef pineal substance in chronic schizophrenia. *New England Journal of Medicine,* 1957, *257,* 919-922.

Anton-Tay, F. Discussion comments in G. E. W. Wolstenholme and J. Knight (Eds.) *The pineal gland.* Edinburgh: Churchill Livingstone, 1971, 363-364.

Bellairs, A. and Carrington, R. *The world of reptiles.* New York: American Elsevier Publishing Company, 1966.

Droscher, V. B. *The magic of the senses.* New York: E. P. Dutton, 1969.

Kappers, J. A. Preface in J. A. Kappers and J. P. Schade (Eds.) *Structure and function of the epiphysis cerebri.* Amsterdam: Elsevier Publishing Company, 1966, ix-xii.

Martini, L. Discussion comments in G. E. W. Wolstenholme and J. Knight (Eds.) *The pineal gland.* Edinburgh: Churchill Livingstone, 1971, 368-372.

Milcu, S. M., Nanu-Ionescu, L. and Milcu, J. The effect of pinealectomy on plasma insulin in rats in G. E. W. Wolstenholme and J. Knight (Eds.) *The pineal gland.* Edinburgh: Churchill Livingstone, 1971, 345-357.

Miline, R. Discussion comments in G. E. W. Wolstenholme and J. Knight (Eds.) *The pineal gland.* Edinburgh: Churchill Livingstone, 1971, 300.

Moszkowska, A., Kordan, C. and Ebels, I. Biochemical fractions and mechanisms involved in the pineal modulation of pituitary gonadotropin release in G. E. W. Wolstenholme and J. Knight (Eds.) *The pineal gland.* Edinburgh: Churchill Livingstone, 1971, 241-255.

Nir, I. Discussion comments in G. E. W. Wolstenholme and J. Knight

(Eds.) *The pineal gland.* Edinburgh: Churchill Livingstone, 1971, 222-225.

Oksche, A. Sensory and glandular elements of the pineal organ in G. E. W. Wolstenholme and J. Knight (Eds.) *The pineal gland.* Edinburgh: Churchill Livingstone, 1971, 127-146.

Reiss, M., Davis, R. H., Sideman, M. B. and Plichta, E. S. Pineal gland and spontaneous activity of rats. *Journal of Endocrinology,* 1963, *28,* 127-128.

Reiss, M., Sideman, M. B. and Plichta, E. S. Spontaneous activity and pineal gland cell density. *Journal of Endocrinology,* 1967, *37,* 475-476.

Reiter, R. J. Comparative physiology: Pineal gland. *Annual Review of Physiology,* 1973, *35,* 305-328.

Tryon, R. C. Individual differences in maze ability. II The determination of individual differences by age, weight, sex and pigmentations. *Journal of Comparative Psychology,* 1931, *12,* 1-22.

Winston, H. D., Lindzey, G., and Conner, J. Albinism and avoidance learning in mice. *Contemporary Research in Behavioral Genetics.* 1967, *63,* 77-81.

Wolley, D. W. and van der Hoeven, T. Prevention of a mental defect of phenylketonuria with seratonin congerers such as melatonin or hydroxytryptophan. *Science,* 1964, *144,* 1593-1594.

Wurtman, R. J. The pineal and endocrine function. *Hospital Practice,* 1969, *4,* 32-37.

Wurtman, R. J. Summary of symposium in G. E. W. Wolstenholme and J. Knight (Eds.) *The pineal gland.* Edinburgh: Churchill Livingstone, 1971a, 379-389.

Wurtman, R. J. Discussion comments in G. E. W. Wolstenholme and J. Knight (Eds.) *The pineal gland.* Edinburgh: Churchill Livingstone, 1971b, 238.

Wurtman, R. J. and Anton-Tay, F. The mammalian pineal as a

neuroendocrine transducer in E. B. Astwood (Ed.) Recent progress in hormonal research: Proceedings of the 1968 Laurention Hormone Conference, Volume 25, New York: Academic Press, 1969, 493-522.

Wurtman, R. J., Axelrod, J. and Kelly, D. E. *The pineal.* New York: Academic Press, 1968.

Part Four

Frontiers of Eye Color Research

Chapter 11

Eye Color, Responsiveness and Sociability

In each of the areas yet to be covered there are some facts to draw on and some old research findings to be reviewed or a new study or two to be presented. However, the fact that the findings in these areas are few and scattered should caution us to keep in mind the tentative nature of our generalizations. One purpose in discussing these areas is to alert other researchers to areas that warrant exploration. In that sense, my purpose here is to generate hypotheses for future research. Another purpose is to place a tentative organization on additional facts about eye color differences that may be helpful in understanding the total phychobiological picture. With those purposes in mind, let us begin by considering the possibility that the eye darkness/reactivity relationship can be expanded to include areas of responsiveness that we have not yet discussed.

If dark eyes are related to a reactive adaptation in nature and light eyes to a nonreactive adaptation, we might expect to find a greater overall sensitivity or responsiveness in the dark-eyed to various environmental stimuli. The dark-eyed may respond to finer or weaker cues which make them sensitive to aspects of the environment missed by the light-eyed. This difference in sensitivity is not totally one-sided, of course. The light-eyed may be more, rather than less, sensitive to certain (e.g., spatial) cues, but the reactive adaptation would seem to require that the dark-eyed individual be acutely responsive to a wide range of environmental stimuli. This generalized responsiveness, if present, might be seen in response to both physical and social stimuli.

Responsiveness to Physical Stimuli

A number of studies have been presented in earlier chapters that show that dark-eyed organisms respond with behavioral reactivity and light-eyed organisms respond with behavioral nonreactivity. This difference, it was suggested, is probably related to differences in physiological responsiveness parallel to behavioral reactivity. There is some support for that position. One measure of autonomic responsiveness, mentioned earlier, is the galvanic skin response (GSR). The GSR measures skin conductance which fluctuates with changes in the level of secretions from the sweat glands. The measured GSR of an individual goes up in response to any stimulus which is strong enough to arouse, stimulate, or startle him. Physiological responsiveness as measured by the GSR appears to be higher in blacks than in whites (Dreger and Miller, 1968) and in Japanese than in whites (Lazarus, Tomita, Opton, and Kodama, 1966). These group differences have been interpreted as probably cultural in origin, but they may reflect genetic differences in autonomic responsiveness among dark-eyed and light-eyed individuals. GSR studies dealing directly with eye color would help to answer that question.

Another source of relevant information is research on responsivity to painful stimuli. Numerous studies have been done to measure the degree to which people experience and respond to pain. Various types of response have been measured: physiological changes, subjective self report of pain, and behavioral reactions such as wincing or pulling away from the painful stimulus.

Several studies having to do with pain have compared different ethnic groups. Chapman (1944) did a study of pain responses to heat radiation applied to the skin in gradually increasing intensity. The subject indicated when he first perceived pain and the experimenter also noted at what point the subject reacted to the pain with facial wincing. The results of the ethnic group comparisons indicated differences as follows:

> Not only did the Negro perceive pain at a lower level, but he also reacted relatively nearer his pain perception than did the Northern European . . . The group of Italians and individuals of Russian Jewish extraction tested had both pain perception and pain reaction values which corresponded more nearly to those of the Negro (Chapman, 1944, p. 5).

A series of studies (Sternbach and Tursky, 1965, Tursky and Sternbach, 1967 and Zborowski, 1969) have been done to compare pain responses of four American ethnic groups: Italians, Jews, Irish, and Old Americans (Anglo-Saxons, usually Protestant). Pain responses of these groups were studied in laboratory experiments and also in hospital situations in which patients from these groups were observed and interviewed in order to gauge their responses to pain occurring as a natural part of disease or treatment. Taking all of the

123

evidence, physiological, subjective report, and behavioral reactions, a clear picture emerges of the Irish and Old Americans as being relatively less responsive and the Italians and Jews as being relatively more responsive to pain. As patients, the Italians and Jews compared to the Irish and Old Americans, expressed pain more openly, complained more, and made more motor responses to pain such as bodily movements, gestures, twisting and jumping. These and similar differences were interpreted by the authors as probably cultural in origin. Undoubtedly, culture does account for some of the differences, but it may not be coincidental that the two more responsive groups are, on the average, darker-eyed than the two less responsive groups.

The same conclusion can be drawn about the research findings of Kenneth M. Woodrow. This investigator measured the degree of pain tolerance by means of a machine that placed pressure on the patient's Achilles tendon in the heel. He found that white patients could tolerate more pain than either black or Oriental patients. Again, there may be cultural explanations to consider, but the differences, as before, indicate that the darker-eyed groups were more responsive to pain than the light-eyed group.

Finally, there is one study to report which was not of ethnic group differences, but a direct comparison of eye color and reaction to pain (Sutton, 1959). The study was done in the Dental School of the University of Melbourne, Australia. More than 400 dental patients were studied in all. All of the patients were Australians of European stock. A high speed drill was used to prepare the teeth of each patient and a four-point scale was used to rate each patient's pain reactions in this situation. If he indicated no pain the lowest rating was given; if he indicated so much pain that a local analgesic was necessary, the highest rating was given and the other two ratings indicated responses between those extremes. In addition to the rating of pain reaction, eye color was also recorded for each patient. The relationship between eye color and pain reaction is shown in Figure 11-1. It can be seen that the darker the eyes of the patient the more likely he was to give a strong pain reaction. This finding is a strong indication that the reactivity dimension related to eye color is one which has autonomic as well as behavioral and perceptual components. It is appropriate, I think, to refer to this general dimension as psychobiological reactivity.

In one of the studies of pain (Zborowski, 1969), it was noticed that Irish and Old American hospital patients responded to pain by withdrawing from other people. This was in contrast to the Jewish

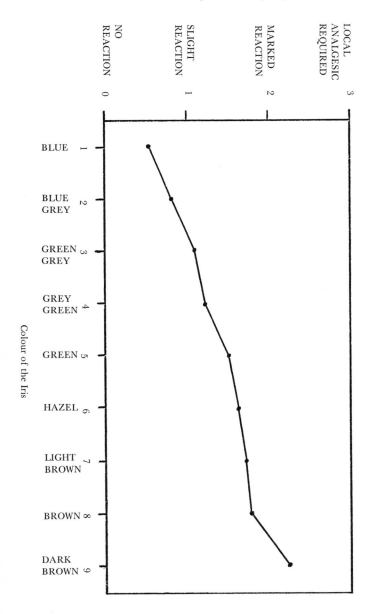

Pain Reaction During Dental Cacity Preparation

Colour of the Iris

Figure 11-1

Association Between the Colour of the Iris and the
Reaction To Pain Resulting From Dental Cavity Preparation
(From Sutton 1959)

and Italian patients who continued to interact socially and communicate their feelings of pain openly and emotionally. This difference in social behavior in a hospital setting may reflect a more general difference in sociability between light-eyed and dark-eyed groups. The evidence, though inconclusive, is worth consideration.

Eye Color and Sociability

Other individuals (especially of the same species) comprise an important stimulus category in the environment of any organism. It is possible that dark-eyed organisms are more socially sensitive and more socially responsive than are light-eyed organisms. As a general statement, that is still a very speculative hypothesis. Animal data on this question have not been systematically gathered except for some that I assembled on the birds of India. A tally was made of all species mentioned as being particularly social or nonsocial. No differences in this regard were found among species of different eye colors. Although that was a clear failure to demonstrate the relationship, other animal studies should be done that differentiate more precisely among various kinds of social behavior. Especially needed are studies that separate social behavior dictated by other needs (e.g., migration or breeding at scarce sites) from social behavior which is more for its own sake.

Among some grasshoppers pigmentation and sociability are very definitely correlated. During a period of dark pigmentation (darker body and darker eyes) the grasshoppers are very gregarious. That period alternates with a period of lighter pigmentation during which the grasshoppers are solitary. These changes in pigmentation and in gregariousness are linked to endocrine changes (Uravov, 1966, p. 378).

Differences in social behavior among various human groups suggest that dark-eyed people are more positively responsive to social stimuli than are light-eyed people. The anthropologist, Edward T. Hall (1966) in his book *The Hidden Dimension* has recorded many examples of how different nationalities structure interpersonal space. Some groups, such as Arabs and Japanese want close contact with other people. Other groups, for example the Germans, like to maintain more interpersonal distance and provide for more privacy. Making the social and spacial adjustment from one culture to another can be difficult:

An American in the Middle East or Latin America is likely to feel crowded and hemmed in—people come too

126

close, lay their hands on him, and crowd against his body. He doesn't feel this in England or Scandanavia where it is the American who perceives the local residents as cold and aloof (Sommer, 1969, p. 69).

Lewis (1972) noted that men in the United States maintain more social distance than do women, but that men of some ethnic groups do not conform to that norm:

> While these sex related stereotypes may be true for large segments of our culture, and are generally reflected in the mass media, they are by no means universal. Among Americans of Italian, Greek, Armenian or Jewish ancestry, for example, there appears to be more physical contact among men. It has been noted that Mexican-Americans, Puerto Ricans, and other Latin Americans tend to stand somewhat closer in casual conversation, generating subtle unease in Anglo-Saxons (Lewis, 1972, p. 57).

These group differences in interpersonal space are viewed by most anthropologists as culturally based and no doubt they often are. Acknowledging culture as a source of learned differences does not, however, explain why the differences originated and were maintained in the first place. One possibility, in this case, is that only those customs of social distance survived which were compatible with the natural inclinations of the group members. Natural inclinations of individuals are, in turn, selected for survival in the environment inhabited by the groups; so, to say that something is cultural is not at all to completely remove it from biological or evolutionary considerations. I consider it noteworthy that the groups described as socially close are dark-eyed and the groups described as socially distant are light-eyed.

Tendencies toward sociability may have evolved in groups to the degree that those tendencies facilitated the individual's efforts to obtain food and escape predators. For animals that employed react-approach-flee tactics the presence of others might have improved chances of success. As soon as any member of the group detected a prey or predator all members of the group could respond. For animals that employed the wait-freeze-stalk tactics the advantages of early detection would seem to have been offset by the obvious difficulties in maintaining effective surprise or concealment while in a group.

Another consideration is that the presence of others tends to increase the physiological activation of the individual. Increased activation, in turn, facilitates performance of "dominant" or uninhibited responses. Increased activation retards, on the other hand, the performance of tasks that require inhibition. Zajonc (1965) has pointed out these relationships and demonstrated that they hold across a wide range of animal and human behaviors. It is clear that the behavioral changes are caused by physiological changes; the presence of others increases the functioning of the adrenal glands which increases general activation. On the other hand, it has been found in a study of rats that isolation decreases the functioning of the adrenal glands and increases the functioning of the pineal gland (Quay, Bennett, Rosenzweig and Krech, 1969).

At the human level, in sports, there are customs which reflect our beliefs about the effect of others on athletic performance. If the athlete is performing a self-paced task (e.g., playing golf or bowling), we, as spectators, if we want him to do well, are as quiet and inconspicuous as possible. If, on the other hand, he is engaged in some reactive task (e.g., boxing or hitting a baseball) we feel that our cheers and yells can only help him. At a basketball game we are quiet only when the play is self-paced (i.e., when a player shoots a foul shot). It would appear, in general, that performance of athletic tasks on which light-eyed do well is hindered by increased activation and the presence of others, and performance of tasks on which dark-eyed do well is facilitated by increased activation and the presence of others.

In addition to audience effects and interpersonal distance, one might also consider receptivity to social influence as a measure of sociability. Karp (1972) has recently completed a doctoral dissertation at Emory University in which he measured the degree to which the opinions of light-eyed and dark-eyed male college students were changed by exposure to the opinions of another (female) student. Dark-eyed students were significantly more influenced, as indicated by change in opinion, than were light-eyed students. The dark-eyed students were also more influenced than were the light-eyed students in their liking for the girl by the degree to which her opinions agreed with their own. A control group of students answered the opinion questionnaire and then answered it again some time later without an influence attempt interspersed between the first and second measures. In that situation, the light-eyed students spontaneously changed their opinions more than did the dark-eyed students. These findings would seem to indicate that dark-eyed people are more responsive than are light-eyed people to social influence, which may be indicative of a greater social sensitivity.

Consistent with that conclusion is evidence which indicates that dark-eyed people are more influenced by social models than are light-eyed people. Gary and Glover (1973) obtained verbal responses from college students after the students had listened to a tape recording of another student performing a similar verbal task. Students who heard a very responsive model gave many responses themselves; students who heard a minimally responsive model tended, themselves, to give few responses. This tendency to be influenced by the behavior of the model was greater for dark-eyed students than for light-eyed students.

If, indeed, there is a higher sensitivity in dark-eyed people to social stimuli, it might be that they are not only more influenced in some situations than are light-eyed people, but are also more influential in those situations that require social sensitivity. For instance, there are many leadership positions in which effectiveness depends more on socio-emotional skills than on the non-personal skills related to the group task. A finding which could be interpreted in such terms resulted from an exploratory study of leadership effectiveness and eye darkness in a situation, high school coaching, which probably depends to a considerable degree on socio-emotional skills.

I should make plain that the study was exploratory and has not been replicated. The study was done primarily because the needed data were already on hand. I had a directory of Texas high school coaches in which there were pictures of some of the coaches together with the school's won-lost records from the year before (Franks, 1967). The pictures were small and some were blurred, but a rater was able to rate the pictures of 63 football coaches on the five-point scale of eye darkness. The ratings were combined to produce two groups: light-eyed coaches (ratings 1, 2 and 3) and dark-eyed coaches (ratings 4 and 5). After the pictures were rated the percentage of games won the year before was computed. A statistical analysis indicated that the dark-eyed coaches had won significantly more of their games than had the light-eyed coaches (Mann Whitney U Test, $Z = 2.34, p < .02$).

Probably as an artifact of the small size and darkness of the pictures, there were only 17 coaches in the light-eyed category and 46 in the dark-eyed category. Of the light eyed coaches in the sample, 41% had won as many or more games the year before than they had lost. This was true for 74% of the dark-eyed coaches in the sample. Turning the comparison around as in Figure 11-2 (and using the 5 point scale of eye darkness) it can be seen that those coaches with darker eyes had, on the average, better records.

129

Of the pictures in the directory that could be rated, there were only 10 of basketball coaches. The six dark-eyed basketball coaches had higher winning percentages than did the four light-eyed coaches and, surprisingly, considering the small sample, the difference was statistically significant (Mann Whitney U Test, U = 1, p = .02).

I must emphasize again that these findings should be considered as very tentative until better controlled studies have been done. However, if these findings prove to be reliable, they will generate numerous questions about eye color and effectiveness in various other interpersonal situations. Based on the limited evidence avail-

Figure 11-2

Eye Darkness and Coaching Success
For A Sample of Texas High School Coaches

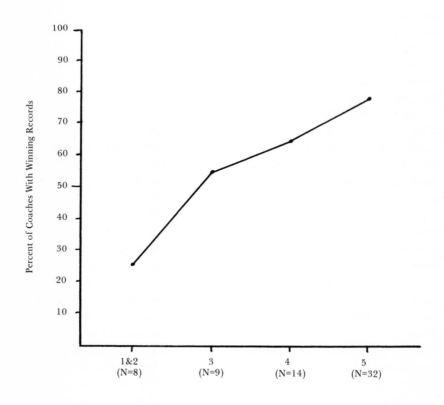

Eye Darkness

able I would hypothesize that dark-eyed people are more effective than light-eyed people in situations that call for social sensitivity and socio-emotional skills.

Another aspect of interpersonal behavior which should be investigated is whether differences in reactivity influence interpersonal attraction. Consistent with differences in reactivity it may be that dark-eyed have more rapid personal tempos than light-eyed. The earlier findings indicative of lower sensory thresholds of response and rapid tempos of responding tend to go together in biological systems. Differences in personal tempo may be an important determinant of interpersonal attraction. We often talk in a metaphorical sense about liking someone because his "vibrations" are like our own or we say that a person's vibrations are different from, but in harmony with, our own. We use such terms to express a general compatibility between our natural or comfortable rhythms (of conversation, bodily movements, etc.) and those of the other person. Human responsiveness to different tempos is certainly subtle and complicated, but not necessarily different in kind from what is seen in other species of animals.

Many animals are drawn to members of the same species because of an inherent attraction for a particular tempo of movement. For instance, butterflies attend to the tempo of wing beats, fireflies to the tempo of flashing, and lizards to the tempo of push-ups done in sexual display. These tempos are species-specific and are crucial in helping the animal find a mate in an environment which may contain many other, slightly different, species (Marler and Hamilton, 1966, p. 384). In other animals the attractive movements may involve an integrated rhythm of various movements or the crucial tempo may be vocal.

In birds, there are wide differences in voice frequency among different species of birds (Armstrong, 1963), it appears that light-eyed birds may have lower voice frequencies than do dark-eyed birds. Herons and owls, for instance, have maximum frequencies below 600, whereas many species of dark-eyed song birds have maximum frequencies above 6000. Perhaps some or all of the eye color differences in voice frequency are accounted for by differences in size, but the subject would appear to warrant investigation.

Recordings of ultrasonic vocalizations in rats indicate that the rat has a low frequency vocalization which accompanies behavioral inhibition and social or sexual withdrawal. It has a high frequency vocalization which accompanies social or sexual readiness and behavioral arousal. These differences in vocalization appear to reflect differences in physiological states (Barfield and Geyer, 1972).

Aspects of voice pattern other than frequency may also be worth studying as possible correlates of eye color. In the Audubon Land Bird Guide (Pough, 1949) information is given about the call patterns of 10 species of vireos. Based on pictures of the ten species, five appear to be relatively light-eyed and five relatively dark-eyed. One aspect of the vireo call is a pause which is very distinct in some species and less pronounced in others. In four of the five lighter-eyed species the pause is reported as very pronounced and in the description of the fifth it is not mentioned one way or the other. In the descriptions of the five dark-eyed species, mention is made in all five that the notes are often run together without distinct pauses. Since the pause part of the call would probably be mediated by some type of physiological inhibition, this difference would appear to be consistent with the overall pattern of differences between light-eyed and dark-eyed organisms. Parenthetically, pauses in human vocalizations (stuttering or more normal pauses) could be systematically recorded and analyzed to determine possible eye color differences. Language differences between tribes or nations of light-eyed and dark-eyed peoples might also be revealing in this regard.

Even assuming that there are eye color differences in tempo and pattern of voice or other socially relevant behaviors the effect on attraction, if any, might be complicated by sex differences. In animal research, there are findings that indicate that eye color and sex can interact to determine attraction and mating behavior. In rats, pigmented males are superior to albinos in both fighting ability and mating success (Parsons, 1967, p. 108). When terriers (dark-eyed) and beagles (relatively light-eyed) are reared together in a common pen, the male terriers have much greater mating success than do the male beagles (James, 1951). Parsons (1967, p. 87) reports a study done by Sturdevant (1915) in which sexual attraction was studied in fruitflies. Some of the flies were of a white-eyed strain and others were of a dark-eyed strain. In the experimental arrangement it was possible to have, first, a situation in which males chose between white-eyed and dark-eyed female mates. Then the situation was changed such that the choice was up to the females. The males, given a choice, tended to choose white-eyed females. The females, given a choice, tended to choose dark-eyed males. In those fruitflies, at least, there was an eye color equivalent of the saying about gentlemen preferring blonds and ladies preferring mates who are tall, *dark* and handsome.

As is the case with all of the other subjects raised in this chapter, we must await further research before we can speak with much confidence about a general relationship between eye darkness and social attraction. The purpose of this chapter has been to raise questions and point out possibilities. The research findings, tentative and scattered though they are, suggest strongly that eye color differences in behavior are not limited to simple motor performance. As a social psychologist, I am especially intrigued by the idea that an inherited physical trait, eye color, may prove helpful in understanding and predicting differences in social behavior.

References

Armstrong, E. A. *A study of birdsong.* New York: Oxford University Press, 1963.

Barfield, R. J. and Geyer, L. A. Sexual behavior: postejaculatory song of the male rat. *Science,* 1972, *176,* 1349-1350.

Chapman, W. P. Measurements of pain sensitivity in normal control subjects and in psychoneurotic patients. *Psychosomatic Medicine,* 1944, *6,* 252-257.

Dreger, R. M. and Miller, K. S. Comparative psychological studies of Negroes and whites in the United States: 1959-1965. *Psychological Bulletin: Monograph Supplement,* 1968, *70,* 1-58.

Franks, R. (Ed.) *1967-68 Texas sports guide of high schools and colleges.* Amarillo, Texas: Ray Franks Publishing Ranch, 1967.

Gary, A. L. and Glover, J. Eye color, sex and creativity. Unpublished manuscript. University of Tennessee, Knoxville, Tennessee, 1973.

Hall, E. T. *The hidden dimension.* Garden City, New York: Doubleday and Company, 1966.

James, W. T. Social organization among dogs of different temperaments, terriers and beagles, reared together. *Journal of Comparative and Physiological Psychology,* 1951, *44,* 71-77.

Karp, E. J. Changes in interpersonal attraction: Effects of physical attractiveness and attitude similarity as measured by the pupilometry technique. Doctoral dissertation, Emory University, 1972.

Lazarus, R. S., Tomita, M., Opton, E., Jr., Kodama, M. A cross-cultural study of stress-reaction patterns of Japan. *Journal of Personality and Social Psychology,* 1966, *4,* 622-633.

Lewis, M. There's no unisex in the nursery. *Psychology Today*. May, 1972, pp. 54-57.

Marler, P. and Hamilton, W. J. *Mechanisms of animal behavior*. New York: John Wiley and Sons, 1966.

Parsons, P. A. *The genetic analysis of behavior*. London: Methuen and Company, 1967.

Pough, R. H. *Audubon land bird guide*. Garden City, New York: Doubleday and Company, 1949.

Quay, W. B., Bennett, E. L., Rosenzweig, M. R. and Krech, D. Effects of isolation and environmental complexity on brain and pineal organ. *Physiology and Behavior*, 1969, *4*, 489-494.

Sommer, R. *Personal space*. Englewood Cliffs, New Jersey: Prentice-Hall, 1969.

Sternbach, R. A. and Tursky, B. Ethnic differences among housewives in psychophysical and skin potential responses to electric shock. *Psychophysiology*. 1965, *1*, 241-246.

Sturdevant, A. H. Experiments on sex recognition and the problem of sexual selection in *Drosophila*. *Journal of Animal Behavior*, 1915, *5*, 351-366.

Sutton, P. R. N. Association between colour of the iris of the eye and reaction to dental pain. *Nature*, 1959, *184*, 122.

Tursky, B. and Sternbach, R. A. Further physiological correlates of ethnic differences in responses to shock. *Psychophysiology*, 1967, *4*, 67-74.

Uravov, B. *Grasshoppers and locusts*. Vol. 1, Cambridge: At the University Press, 1966.

Zajonc, R. B. Social facilitation. *Science*, 1965 *149*, 269-274.

Zborowski, M. *People in pain*. San Francisco: Jossey-Bass, 1969.

Chapter 12

Eye Color, Physical Size and Physical Disorders

If eye color proves to be important to the functioning of the photo-neuro-endocrine system, as seems likely, future research can be expected to turn up many additional correlates of eye color. As more is learned about the functions of the pineal gland, the hypothalamus, and other organs affected by retinal sensitivity to light, we will know better what relationships to look for. Already there are numerous possible hypotheses that could be derived from the material covered in Chapters Nine and Ten. Additional hypotheses might be advanced based on generalizations from eye color findings that have already been made. Since so little eye color research has been done the possible directions are many. Two of the many promising areas for future eye color research are physical size and physical disorders.

Physical Size

Based on casual observation, and on the average, light-eyed human populations seem to be taller than dark-eyed human populations. There are, of course, many dark-eyed people that are very tall and many light-eyed people that are short, but we are talking here only of average height throughout the world. Even though height is inherited we cannot be sure that observed differences are entirely genetic since many factors influence height. Diet, other environmental factors, and genetic factors such as amount of inbreeding versus outbreeding, have to be considered. On the surface, though, there does appear to be an inverse relationship between the eye darkness and height of different human populations. A similar pattern was found in data I have collected on eye darkness and size in birds. First, all species of birds in America (457 species, from Pearson, 1936) were considered. For each species, length and eye color were recorded. Eye color was rated, as before, on the five point scale of darkness. Figure 12-1 shows median length for each level of eye darkness. The tendency was clearly for darker-eyed birds to be shorter in length.

Since some families of birds had many species and others only a few, I decided to test the relationship in an additional manner. For each family of American birds I computed average eye darkness and average length and then correlated the two figures. In this analysis an entire family of species served as just one case. The correlation between the two figures, average eye darkness and average length, was negative and statistically significant (Product Moment Correlation $r = .432, N = 66, p < .001$).

Yet another analysis was done of the data on American birds. This was a within-family matched comparison based only on the 28 (of 66) families which contained both light-eyed and dark-eyed species. In this comparison, the light-eyed birds were larger than the dark-eyed birds but not significantly so (Matched t test, $t = .859$, 28 cases, $p < .05$).

I replicated these analyses with all species of birds of India (308 species, from Kinnear, 1949). Median length in this group also was greater for light-eyed than dark-eyed species as shown in Fugure 12-2. The correlation of average eye darkness by average length for families of birds was negative and statistically significant ($r = -.471$, 62 cases, $p. < .001$). The within-family matched comparison was, as before, directional but not significant (Matched $t = .961$, 24 cases, $p. < .05$).

Just to be sure, I replicated it one more time with species of birds from the Southern Hemisphere, birds of New Guinea (603 species— from Rand and Gilliard, 1968). Again, median length was greater in lighter eyed birds as shown in Figure 12-3. Again, average eye darkness and average length were negatively correlated for families of birds ($r = -.383$, 70 cases, $p < .01$). Also as before, the within-family comparison was directional but not statistically significant (Matched $t = 1.052$, 26 cases, $p < .05$).

For birds, at least, eye color and size are clearly related with light-eyed species being larger than dark-eyed species.

If size is related to eye color, then what about other things that tend to be related to size such as growth rate? Fast growth is positively related to small size. This knowledge is now being put to use in preventing a problem anticipated by some young girls who fear

Figure 12-1

EYE DARKNESS AND MEDIAN LENGTH
OF 457 SPECIES OF BIRDS OF AMERICA

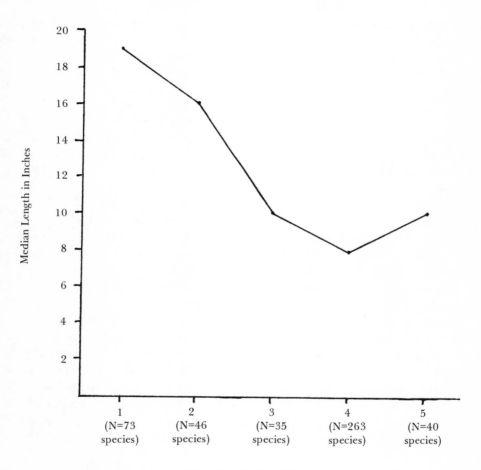

Figure 12-2

EYE DARKNESS AND MEDIAN LENGTH
OF 308 SPECIES OF BIRDS OF INDIA

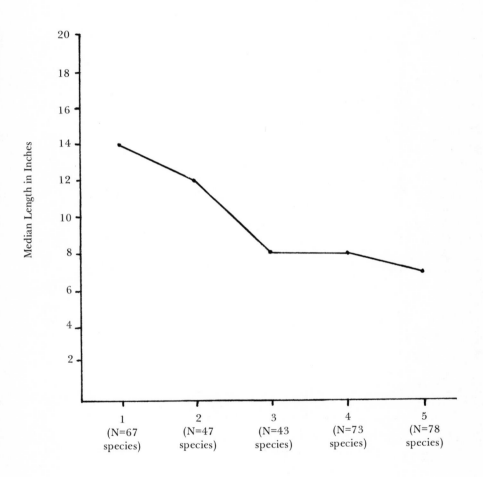

Figure 12-3

EYE DARKNESS AND MEDIAN LENGTH
OF 603 SPECIES OF BIRDS OF NEW GUINEA

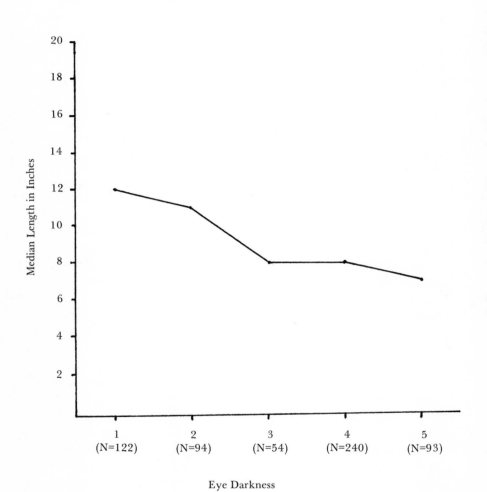

becoming too tall. Hormonal treatments are being used experimentally to speed growth rate and thereby reduce adult height. Investigation of eye color and growth rate would seem to be a fruitful area for study.

Physical Disorders

Our understanding of normal functions is often facilitated by the study of deviant or abnormal functions. An understanding of the relationship of eye color to biological variables may be aided by attention to medical research. Perhaps, this understanding, in turn, will provide additional knowledge that could be useful in medical research.

Among disorders that might be related to eye color are obesity and diabetes. The light-influenced endocrine regulators may be implicated in these disorders. The pineal gland is involved in carbohydrate metabolism and appears to have an insulin-like effect. Pinealectomized animals, for instance, have been reported to exhibit diabetes-like symptoms. It is also known that animals with lesions in the hypothalamus overeat and become very obese. Likewise, people who are blind have a tendency to be obese and differ from non-blind people on carbohydrate balance and insulin balance. Each of these facts suggests the possibility that eye darkness, if related to light input, could be correlated with disorders of body metabolism.

There are also hints of such relationships that come from other sources. In reading about different species and families of birds, I noted that many examples seemed to be present of dark-eyed birds that were prone to become plump when food was abundant. Results of racial comparisons of humans are also suggestive. Predisposition for obesity and diabetes appears to be higher in blacks than in whites (Dreger and Miller, 1968), and higher in American Indians than in whites (Goodall, 1972). Likewise, there are high rates of diabetes among Jews and Hindus (Allen, 1959).

Finally, the work of Schachter (1971) on obese humans is also suggestive. Schachter has pointed out a parallel between the behavior of obese humans and the behavior of rats made obese by means of hypothalamic lesions. Many interesting studies were reported that were interpreted as indicating that the obese, whether human or rat, are more "stimulus bound" or "reactive". The obese rats, for instance, show less "freezing" behavior and are better than normal rats at learning active avoidance, but are poorer than normal rats at learning passive avoidance. Among humans, the obese are faster than the non-obese on perceptual motor tasks, but their performance is more subject to disruption from distracting stimuli. Many such examples of rat behavior and human behavior were given by Schachter to support the idea that the obese are more reactive than the non-obese. Only more research can determine whether the reactivity noted by Schachter as characteristic of the obese is related to the reactivity which I have found characteristic of the dark-eyed.

Other areas that should be investigated are alcoholism and responses to various drugs. If dark eyes are positively related to physiological sensitivity this might be seen in a greater response to various drugs including alcohol. A greater or earlier response to alcohol might serve to decrease the amount of alcohol consumed by the steady drinker since a smaller amount would result in an equivalent effect. For that reason, persons who are most sensitive to alcohol would, perhaps, be less likely to drink enough to become physiologically addicted. That this relationship may explain the low level of alcoholism among Orientals has been suggested by Wolff (1972). He found that Orientals were more physiologically responsive to alcohol than were Caucasians. This relationship was demonstrated in adults and also in newborn babies and was interpreted as probably genetic and related to "variations in autonomic reactivity".

A researcher in Chile (Cruz-Coke, 1970) has reported that alcoholism is more common among heavy drinkers of European descent than among heavy drinkers of American Indian descent. He has likewise reported a positive relationship between inherited color-blindness and an inherited predisposition to alcoholism. This finding has been challenged by a researcher in the United States (Smith, 1972) who believes, based on his research, that the differences in color perception are only a by-product of excessive drinking. The reason this question is relevant to the present discussion is that light-eyed and dark-eyed people differ in color vision. A hypothesis which should be tested is that light-eyed people are less responsive physiologically than dark-eyed people to alcohol and similar drugs.

Another area for future research involves blood disorders. I have been told informally that surgeons have observed that light-eyed people bleed more than dark-eyed people. Added to that observation is the fact that an inherited disease characterized by excessive bleeding (hemophilia) is unusually common in Caucasian males (Ratnoff, 1966). Excessive bleeding, likewise, is associated with albinism (Witkop, 1971).

In considering possible relationships between eye color and

142

various disorders attention should be paid not only to light-eyed and dark-eyed, but also those eye colors intermediate in darkness. Based on observations made at a Canadian hospital, it has been suggested that persons with eye colors in the midrange differ from other people in being unusually variable from one time to another on physiological measures (Kent, 1956) and having a higher than average incidence of schizophrenia (Kent, Christie, Tunis, Lehman and Cleghorn, 1956).

Other disorders could be added to the list, but the ones given seem unusually worthy of research since for most, there are several reasons to suggest a possible relationship with eye color. One should keep in mind, however, that physical disorders are very complex. A particular disorder may come in different forms with different genetic causes for each. Eye darkness could be positively related to one cause and unrelated or even negatively related to some other cause.

A case in point is diabetes. Clinical reports relating eye color to diabetes have found relationships which are consistent within a family (i.e., children of one eye color inherit the disease and children of the other eye color do not), but which eye color is linked to the disease is not always the same. In some families with a known history of diabetes it is only the light-eyed who get the disease and in other families it is only the dark-eyed who get the disease. Since there are different types of diabetes there can be various genetic linkages. What seems to be clearly indicated though is that some of the genes for eye color are associated with genes involved in diabetes (Gates, 1946, p. 548).

In addition to specific disorders as a point of origin for medically-relevant research, normal differences on physiological measures should also be investigated. Information based on physiological responses to lights of different wavelengths provides clues to the measures that should be checked. Gerard (1958) had people sit in a dark room and look at a screen which was lighted by a projecter equipped with colored filters. Watching a blue light had physiological effects that differed from those of watching a red light. Red light, as compared with blue light resulted in higher systolic blood pressure, higher palmer skin conductance, higher respiration rate, higher frequency of eyeblinks and fewer alpha waves in the visual cortex of the brain. To the degree that ocular pigmentation acts as a red filter we might find similar differences as a function of eye darkness.

143

References

Allen, F. M. Diabetes mellitus in *Encyclopedia Americana, 1959 Edition, Volume IX* New York: Americana Corporation, 1959, 54-56.

Cruz-Coke, R. *Color blindness, an evolutionary approach.* Springfield, Illinois: Charles C. Thomas, 1970.

Dreger, R. M. and Miller, K. S. Comparative psychological studies of Negroes and whites in the United States: 1959-1965 *Psychological Bulletin: Monograph Supplement,* 1968, *70,* No. 3, Part 2, 1-58.

Gates, R. R. *Human genetics.* New York: MacMillan, 1946.

Gerard, R. M. Differential effects of colored lights on psycho-physiological functions. Doctoral Dissertation, University of California at Los Angeles, 1958.

Goodall, K. Tie line: A metabolic clue to Indian endurance and intolerance for alcohol. *Psychology Today,* 1972, *6,* 16.

Kent, I. Human iris pigment. I—A concept of individual reactivity with implications in health and disease. *Canadian Psychiatric Association Journal,* 1956, *1,* 99-106.

Kent, I., Christie, R. G., Tunis, M. M., Lehman, H. E. and Cleghorn, R. A. Human iris pigment. II—Factors in schizophrenia. *Canadian Psychiatric Association Journal,* 1956, *1,* 107-108.

Kinnear, N. B. *Whistler's popular handbook of Indian birds.* Edinburgh: Oliver and Boud, 1949.

Pearson, T. G. (Ed.) *Birds of america.* Garden City, New York: Garden City Publishing Company, 1936.

Rand, A. L. and Gilliard, E. T. *Handbook of New Guinea birds.* Garden City, New York: The Natural History Press, 1968.

Ratnoff, O. D. Hereditary disorders of hemostasis in J. B. Stanbury, J. B. Wyngaarden and D. S. Fredrickson (Eds.) *The metabolic basis of inherited disease.* New York: McGraw-Hill Book Company, 1966, 1137-1175.

Schachter, S. Some extraordinary facts about obese humans and rats. *American Psychologist,* 1971, 36, 129-144.

Smith, J. W. Color vision in alcoholics. *Annals of the New York Academy of Sciences,* 1972, *197,* 143-147.

Witkop, C. J. Albinism in H. Harris and K. Hirschhorn (Eds.) *Advances in human genetics.* Volume II, New York: Plenum Press, 1971, 61-142.

Wolff, P. H. Ethnic differences in alcohol sensitivity. *Science.* 1972, *175,* 449-450.

Chapter 13

Sex Differences and Eye Color Differences

In exploring eye color differences, I became aware that eye color differences seemed to be related to sex differences. The eye color differences and the sex differences often fell into an interesting pattern. On those psychobiological dimensions on which light-eyed tended to be higher than dark-eyed, males tended to be higher than females. On those dimensions on which dark-eyed tended to be higher than light-eyed, females tended often to be higher than males. By no means was this always the case, but it was noted frequently.

If these tendencies are, as I believe, more than coincidental, we have yet another guide to use in searching for additional eye color differences. More important, perhaps, the parallel between eye darkness differences and sex differences provides a hint about the genetics involved in at least some psychobiological differences related to eye color.

In this chapter, the evidence which suggested the parallel will first be presented and that will be followed by some notions of a theoretical nature about what might be involved genetically.

The Eye Color and Sex Parallel

Some eye color and sex differences were noted in the area of vision. Light-eyed people are apparently better than dark-eyed people at making certain perceptual distinctions involved in spatial-perceptual abilities (Jahoda, 1971) and males consistently score higher than females on tests of spatial-perceptual ability (Tyler, 1965, p. 245). Spatial-perceptual ability will be dealt with more fully later in the chapter.

There is also a parallel in color preference. Several lines of evidence support the idea that dark-eyed organisms respond relatively more to colors at the red end of the spectrum and light-eyed organisms respond relatively more to colors at the blue end of the spectrum. In birds, as well as in many other animals, including fish, it is the male which has the brighter red and yellow colors. If these colors have evolved as a result of their advantage in attracting members of the opposite sex, the inference to be drawn is that females are more responsive to the bright colors than are males. It has likewise been reported for humans that "among men blue is the first choice and red is the favorite of women" (Cheskin, 1948, p. 61).

Many lines of evidence were drawn to demonstrate that among humans and animals, in a variety of situations, dark-eyed are more behaviorally reactive than are light-eyed. Most of these comparisons, involving sports performance and animal behavior in natural settings, did not allow for investigation of sex differences. In one area of eye color difference related to reactivity, speed of locomotion, the parallel does not hold, or at least it does not hold for human beings. Human males run faster than human females. Differences in anatomy are probably more important in this case than are differences in physiology.

In addition to the sports data and observations from natural settings, results from some laboratory experiments with mammals were also mentioned as supporting the hypothesis that dark-eyed are more reactive than light-eyed. In several such instances we do have informaion on sex differences. Mahut (1958) studied fearfulness in different breeds of dogs. I pointed out earlier that, based on her data, more light-eyed dogs than dark-eyed dogs were hesitant and reluctant to approach a novel object. Mahut reported that for two breeds there was a sex difference in approach behavior such that males when approaching a strange object were more likely than females to exhibit "stalking" behavior.

Another laboratory study (Tryon, 1931) involved maze learning in rats. It was found that rats with lightly pigmented eyes performed better than rats with darkly pigmented eyes. The reason given for this difference was that the light-eyed rats were less "reactive" to extraneous stimuli. Tryon also reported that male rats did significantly better on the maze task than did female rats.

Another study (Winston, Lindzey and Connor, 1967) with rodents, mice in this case, was also reported. On an avoidance learning task, it was found that pigmented mice learned to avoid the average situation in an active manner, but that albino mice learned to avoid the situation in a passive manner. When the only way to avoid was to do so actively, the pigmented mice were clearly superior to the albino mice. Likewise, it has been found that females learn active avoidance better than males (Beatty and Beatty, 1970).

Greater reactivity in dark-eyed than in light-eyed humans was indicated by behavioral and psysiological responses to painful stimuli and by greater GSR (Galvanic Skin Response) magnitude. It has likewise been found that human females have lower pain thresholds than human males and human females have higher GSR's than males (Kopacz and Smith, 1971). In addition, female rats have lower flinch and jump thresholds to shock than do male rats (Beatty and Beatty, 1970). The fact that females learn classically conditioned responses faster than males has been attributed to the females higher level of "emotional reactivity" (Spence and Spence, 1966). In a review of human sex differences in physiological activation, Duffy (1962, p. 225) reports that most measures of arousal show no difference between the sexes or show a difference in the direction of a higher level of activation on the part of females.

Greater activation would result, in general, in greater responsiveness. For example, in seeing-eye dogs, female dogs are reported to be more sensitive than male dogs to both auditory and tactile stimuli (Adsell, 1966, p. 172).

In addition to evidence for differences in responsiveness to physical stimuli, some evidence was also presented to suggest that dark-eyed people are more responsive to social stimuli than are light-eyed people. Tyler (1965, p. 259) reviews studies that indicate that females are more sociable, more suggestible and more accurate in measures of social perception than are males. She concludes, based on the available data that females have a more personal or social orientation to life than do males.

Consistent with that conclusion, Gary and Glover (1973) in their study of modeling effects found not only that dark-eyed were more influenced by the behavior of another person than were light-eyed, but also that females were more influenced than were males.

Some evidence was presented to indicate that light-eyed organisms tend to be larger than dark-eyed organisms and it was suggested, consistent with that, that the light-eyed might also have a slower rate of growth. It is not true in all animals (e.g., hawks), but in many species, including humans, the male is larger than the female. Among humans, at least, males also have a slower rate of growth than do females and are later in reaching their adult level of development (Ghent, 1961).

Finally, the suggested eye color differences in physical disorders, though very speculative also reflect the same eye color and sex parallel. For example, more males than females are afflicted with hemophilia (Ratnoff, 1966) and alcoholism (Winokur and Clayton, 1968).

On the other hand, dark-eyed individuals are perhaps more susceptible to diabetes and obesity than are the light-eyed. Consistent with the parallel, more women than men have diabetes (Allen, 1959) and there are sex differences present at birth in taste preferences which are related to obesity. Nisbett and Gurwitz (1970) studied responsiveness to sweet taste in infants and found a greater responsiveness in obese than in non-obese babies and a greater responsiveness in female babies than in male babies. They noted that similar differences are found in infants and adults and among animals other than humans. They suggest the possibility of congenital sex differences in eating behavior and state; "There is reason to suspect that such differences may exist, with females showing behavior characteristic of obese organisms and males showing behavior characteristic of non-obese organisms" (p. 246).

To summarize, some suggestive evidence of a parallel between eye color differences and sex differences on the reactivity dimension has been observed on a number of psychobiological variables. On those dimensions, dark-eyed and females tend to respond alike and light-eyed and males tend to respond alike. It should also be remembered that females have slightly darker eyes than do males. That fact may play a part in determining the parallel, but the size of the sex differences are great enough to make it unlikely that that is the full explanation.

A Simple Model of Sex and Eye Color Differences

It appears that both eye color and sex have effects on general psychobiological reactivity. By assuming that these effects are additive, we can generate a simple model as follows:

Theoretical Levels of Psychobiological Reactivity

	Males	Females
Dark-eyed	Medium	High
Light-eyed	Low	Medium

Dark-eyed females, according to the model, would be the most reactive and light-eyed males the least reactive. Light-eyed females and dark-eyed males would be intermediate in reactivity. The model was presented in terms of reactivity rather than non-reactivity, but predictions from the model apply equally well to reactive or nonreactive behaviors or tasks (i.e., on nonreactive tasks it is predicted that light-eyed males should score highest and dark-eyed females score lowest). An ideal research design for testing reactivity and nonreactivity would be one in which both types of scores are obtained and a comparison made in relative as well as absolute terms.

Data obtained by a young colleague, Jordan (1972), can be used to illustrate an area in which the model has predictive value. Jordan tested the effects of sex and eye color (blue-eyed and brown-eyed Caucasians) on various tasks. Six cognitive tasks were chosen specifically because they were believed to be ones on which sex differences should occur. The tasks chosen were selected to be consistent with the dimensions suggested by Broverman, Klaiber, Kobayashi and Vogel (1968). These authors have suggested that females perform better on simple perceptual motor tasks that require speed of response. They suggested that males perform better on inhibitory perceptual-restructuring tasks on which the initial responses are likely to be wrong.

Although others in interpreting sex differences have stressed the possible effects of learning and sex roles, Broverman and his co-workers made a strong case for their hypothesis that sex differences in cognitive abilities follow from sex-related differences in physiology. They support their position with findings from animal and human research and show how sex differences in endocrine functioning could account for the observed behavioral or cognitive differences. They conclude that females have a higher level of activation which leads to superior performance on perceptual motor tasks; they also conclude that the male's lower level of activation leads to superior performance on inhibitory perceptual-restructuring tasks.

Of the tests administered by Jordan to blue-eyed and brown-eyed male and female Caucasian college students, three tests were chosen to measure perceptual-motor skills and three were chosen to measure perceptual-restructuring skills. The perceptual-motor tests were considered to be dependent on "reactive" skills and it was expected, therefore, that on these measures the brown-eyed would score better than blue-eyed, and females better than males. The perceptual-restructuring tasks were considered nonreactive and it was expected, therefore, that blue-eyed would score better than brown-eyed and males score better than females.

The perceptual-motor tests consisted of (1) speed and accuracy of finding and canceling specific letters on a printed page, (2) speed of naming 100 color patches and (3) the Digit Symbol subtest of the Wechsler Adult Intelligence Scale which requires speed and accuracy in using a set of novel symbols. All three of the tasks demand perceptual-motor speed. On each of the three tasks females scored higher than males and brown-eyed scored higher than blue-eyed. (The differences on individual tasks were small and not statistically significant, however the consistency of the pattern across tasks was statistically significant.) The pattern of results was exactly the same for each of the three tasks with the groups ordered from best to poorest performance as follows: (1) Brown-eyed females, (2) Blue-eyed females, (3) Brown-eyed males and (4) Blue-eyed males.

The perceptual-restructuring tasks consisted of (1) a mirror-maze task on which the person must trace a simple paper and pencil maze while viewing the maze only by means of a mirror, (2) the Group Embedded Figures Test which requires finding of specific figures embedded in a mass of lines, (3) the Object Assembly subtest of the Wechsler Adult Intelligence Scale which requires rapid solving of form puzzles. All of the tasks require inhibition of wrong responses. Unlike the consistent results found for perceptual-motor tasks the results for the various perceptual-restructuring tasks were not consistent. The fact that the tasks required speed as well as restructuring may have made them less discriminating than they would have been otherwise. At any rate, the results were not consistent from one test to another and neither sex nor eye color functioned as a strong predictor. Only by combining the results of the three tests does a slight pattern emerge. The pattern is consistent with expectations in that average performance of the sex-by-eye color groups on these tasks tended to be opposite from what was found on the perceptual-motor tasks. Thus, on the perceptual-restructuring tasks, brown-eyed females did poorest and blue-eyed males did best; however, this difference did not approach statistical significance.

In Figure 13-1 the curves are shown for performance on the two types of tasks as found by Jordan. The contrast in performance on the two types of tasks is in agreement with the theoretical model. The results suggest differential cognitive abilities as a function of sex and eye color rather than general superiority for either sex or eye color group.

A similar picture of the contrast between perceptual-motor speed and spatial-perceptual ability can be seen from scores for different ethnic groups. These scores provide additional support for a parallel

151

between sex and eye color differences. Backman (1972) has published information on patterns of abilities for males and females of four American ethnic groups: Negroes, Orientals, Jews and Non-Jewish Whites. She analyzed various ability scores of 12,925 12th grade students who were all from upper middle class or lower middle class homes. Two of the ability factors reported were perceptual measures that contrast with each other on the perceptual-motor versus perceptual-restructuring dimension. The first of these factors was described as "Perceptual Speed and Accuracy, a measure of visual-motor coordination under speeded conditions" and the second was described as "Visual Reasoning, a measure of reasoning with spatial forms."

We can assume that the non-Jewish white group was lighter-eyed than any of the other three ethnic groups. It is significant that the non-Jewish whites had the lowest average scores of any group on perceptual speed and accuracy and the highest average scores on the measure of spatial ability. Likewise, for every ethnic group, females scored higher than males on perceptual speed and accuracy and males scored higher than females on the measure of spatial ability.

Considering available evidence, there appears to be some characteristic, probably genetic, which, if present, causes a person to do well on tasks that require spatial-restructuring ability and to do poorly on tasks that require perceptual-motor speed. The evidence indicates that males and light-eyed are more likely to have that characteristic than are females and dark-eyed.

Some Genetic Considerations

In an earlier part of this book (Chapters 9 and 10) I tried to suggest physiological mediators of the relationship between eye color and behavior. I am not a physiologist and that research will have to be done by persons more qualified in that area than I. However, I did feel that, having concentrated on eye color as such, I had a unique perspective from which to speculate about what type of physiological functioning might be involved. I wish, in the same spirit, to make a few remarks about the possible genetics involved.

The fact that there are sex differences in reactivity points to possible sex-linked inheritance. Males differ from females in having only one X chromosome rather than two. In sex-linked traits, males exhibit phenotypically any characteristic inherited on the X chromosomes. Therefore, any characteristic which is recessive and sex-linked is more common in males than in females.

There are several reasons to believe that one gene involved in psychobiological reactivity is carried on the X chromosome. First are the sex differences as already noted. There are also indications of X chromosome involvement that are more directly related to eye color. Eye color, itself, is polygenic, but there is reason to believe that at least one of the genes involved is sex-linked and carried on the X chromosome (Moody, 1967, p. 174). This gene may be linked to other X recessive genes that help determine level of reactivity.

There is evidence that various endocrine functions and dysfunctions are influenced by sex-linked genes (Goodman, 1970). A similar hint of possible X chromosome involvement in endocrine functions related to light is provided by the recent finding that, in fruitflies, biological clock functions are influenced by a gene found on the X chromosome (Konapaka and Benzer, 1971).

Stafford (1961) has hypothesized that spatial ability is transmitted by a recessive gene carried on the X chromosome. Garron (1970) reviewed evidence which, for the most part, supports Stafford's position. It was noted, for instance, that, for boys, their spatial ability scores correlate with scores made by their mothers but not those made by their fathers; this indicates sex-linked inheritance of spatial abilities since a boy gets his one X chromosome from his mother. Spatial ability scores for girls do correlate with their father's scores and this is consistent with the hypothesis since girls get one of their two X chromosomes from their father.

It seems a reasonable possibility that the same genes are accounting for increased spatial-perceptual ability and decreased perceptual-motor speed. A further possibility is that these sex-linked genes are responsible for many of the other observed sex and eye color differences in behavioral reactivity.

Much more research will need to be done before we can have a fully integrated theory which incorporates behavioral, physiological and genetic facts about eye color. However, a start has been made in that direction and the prospects look very promising.

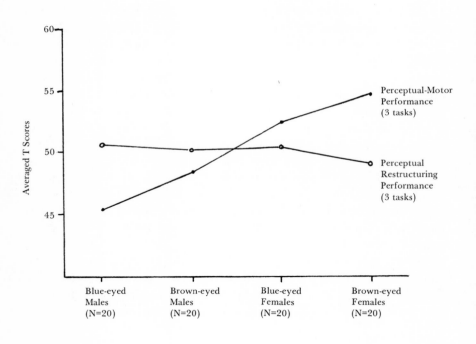

Figure 13-1

PERFORMANCE OF CAUCASIAN COLLEGE STUDENTS
ON PERCEPTUAL-MOTOR TASKS AND
PERCEPTUAL-RESTRUCTURING TASKS AS A FUNCTION
OF SEX AND EYE COLOR (Adapted from Jordan, 1972)

Perceptual-Motor
Performance
(3 tasks)

Perceptual
Restructuring
Performance
(3 tasks)

Averaged T Scores

Blue-eyed
Males
(N=20)

Brown-eyed
Males
(N=20)

Blue-eyed
Females
(N=20)

Brown-eyed
Females
(N=20)

Summary Statement

In this last section, Part Four, we have considered a number of areas that may prove profitable for future eye color research. What we already know leads us to believe that these areas are ones in which eye color differences are related to other individual differences. As we gather more facts we should be able to integrate what is known about physical traits, such as eye color, sex, size, etc., and what is known about behavioral traits, such as sensitivity, reactivity, sociability, etc., into a comprehensive theory of individual differences.

As we learn more about the variety of meaningful individual and group differences, we may come to appreciate anew the degree to which each of us represents a unique combination of inherited and learned traits.

References

Allen, F. M. Diabetes mellitus in *Encyclopedia Americana, 1959 Edition, Volume IX.* New York: Americana Corporation, 1959, 54-56.

Asdell, S. A. *Dog breeding.* Boston: Little, Brown and Company, 1966.

Backman, M. E. Patterns of mental abilities: Ethnic, socioeconomic, and sex differences. *American Educational Research Journal,* 1972, *9,* 1-12.

Beatty, W. W. and Beatty, P. A. Hormonal determinants of sex differences in avoidance behavior and recativity to electric shock in the rat. *Journal of Comparative and Psysiological Psychology,* 1970, *73,* 446-455.

Broverman, D. M., Klaiber, E. L., Kobayashi, Y. and Vogel, W. Roles of activation and inhibition in sex differences in cognitive abilities. *Psychological Review,* 1968, *75,* 23-50.

Cheskin, L. *Colors.* New York: Liveright, 1948.

Cruz-Coke, R. *Color blindness: An evolutionary approach.* Springfield, Illinois: Charles C. Thomas, Publisher, 1970.

Duffy, E. *Activation and behavior.* New York: John Wiley and Sons, Inc., 1962.

Garron, D. C. Sex-linked, recessive inheritance of spatial and numerical abilities, and Turner's syndrome. *Psychological Review,* 1970, *77,* 147-152.

Gary, A. L. and Glover, J. Eye color, sex and creativity. Unpublished manuscript. University of Tennessee, Knoxville, Tennessee, 1973.

Ghent, L. Developmental changes in tactile thresholds on dominant and nondominant sides. *Journal of Comparative and Psysiological Psychology,* 1961, *54,* 670-673.

Goodman, R. M. The family pedigree and genetic counseling in R. M. Goodman (Ed.) *Genetic disorders of man*. Boston: Little, Brown and Company, 1970, 85-104.

Jahoda, G. Retinal pigmentation, illusion susceptibility and space perception. *International Journal of Psychology*, 1971, *6*, 199-208.

Jordan, J. J. III *An investigation of sex, eye darkness and social class differences in perceptual motor and cognitive abilities*. Doctoral Dissertation, Georgia State University, 1972.

Konapaka, R. J. and Benzer, S. Clock mutants of *Drosophila melanogaster*. *Proceedings of the National Academy of Science*, U.S.A., 1971, *68*, 2112-2116.

Kopacz, F. M. and Smith, B. D. Sex differences in skin conductance measures as a function of shock threat. *Psychophysiology*, 1971, *8*, 293-303.

Mahut, H. Breed differences in the dog's emotional behavior. *Canadian Journal of Psychology*, 1958, *12*, 35-44.

Moody, P. A. *Genetics of man*. New York: W. W. Norton and Company, 1967.

Nisbett, R. E. and Gurwitz, S. B. Weight, sex and the eating behavior of human newborns. *Journal of Comparative and Psysiological Psychology*, 1970, *73*, 245-253.

Ratnoff, O. D. Hereditary disorders of hemostasis in Stanbury, J. B., Wyngaarden, J. R., and Fredrickson, D. S. *The metabolic basis of inherited disease*, New York: McGraw-Hill Book Company, 1966, 1137-1175.

Spence, K. W. and Spence, J. T. Sex and anxiety differences in eyelid conditioning. *Psychological Bulletin*, 1966, *65*, 137-142.

Stafford, R. E. Sex differences in spatial visualization as evidence of sex linked inheritance. *Perceptual and Motor Skills*, 1961, *13*, 428.

Tryon, R. C. Individual differences in maze ability. II The Determination of individual differences by age, weight, sex and pigmentation. *Journal of Comparative Psychology*, 1931, *12,* 1-22.

Tyler, L. E. *The psychology of human differences.* New York: Appleton-Century-Crofts, 1965.

Winokur, G. and Clayton, P. J. Family history studies: IV Comparison of male and female alcoholics. *Quarterly Journal of Studies on Alcohol,* 1968, *29,* 885-891.

Winston, H. D., Lindzey, G. and Connor, J. Albinism and avoidance learning in mice. *Contemporary Research in Behavior Genetics.* 1967, *63,* 77-81.

BIOGRAPHICAL SKETCH

Morgan Worthy grew up in Greenville, South Carolina, where he earned undergraduate degrees from North Greenville College and Furman University. After spending four years in the Air Force as a communications analyst and one year as a high school history teacher, he enrolled as a graduate student at the University of Florida. In 1965, he graduated with an M.A. and a Ph.D. in social psychology. He later spent a year in post doctoral training in clinical psychology at the Veterans Hospital, Hampton, Virginia.

For the last seven years, Dr. Worthy has been on the faculty of Georgia State University. His duties have included teaching, research, counseling, and supervision of graduate students. His research findings have been published in a number of scientific journals. He is a member of the American Psychological Association and the American Association for the Advancement of Science.

His original interest in eye color grew out of two other long-standing interests: sports and individual differences.

In addition to his research in eye color his recent interests include the development of ways to teach flexible or creative thinking. With that purpose in mind, he has written a book of puzzles, called *Crossmatch Creatagrams,* that will be published soon by Droke House/Hallux.

DATE DUE